T5-DGT-933

# AN INTRODUCTION

## TO

## ORGANIC CHEMISTRY

# AN INTRODUCTION TO ORGANIC CHEMISTRY

JOHN CARNDUFF

*Chemistry Department,*
*The University of Glasgow,*
*Glasgow, Scotland*

JOHN WILEY & SONS

Chichester · New York · Brisbane · Toronto

FERNALD LIBRARY
COLBY-SAWYER COLLEGE
NEW LONDON, N.H. 03257

QD
251.2
C 37

83039

Copyright © 1978 by John Wiley & Sons, Ltd.

All rights reserved.

No part of this book may be reproduced by any
means, nor transmitted, nor translated into a
machine language without the written permission of
the publisher.

*Library of Congress Cataloging in Publication Data:*

Carnduff, John.
  An introduction to organic chemistry.
  Includes index.
  1. Chemistry, Organic. I. Title.
QD251.2.C37      547       78-16664
ISBN 0 471 99647 5
ISBN 0 471 99685 8 pbk.

Printed in Great Britain by Spottiswoode Ballantyne Ltd. Colchester and London

For Ann

# Acknowledgements

I am grateful to many people who helped me prepare this book. My colleagues Professor D. W. A. Sharp and Dr. A. H. Johnstone encouraged me to write it, gave advice on style and content and made detailed comments on the entire manuscript. Numerous teachers in schools and universities have made comments and criticisms which have been of value. In particular Professor J. M. Tedder (St. Andrews University) read the entire manuscript critically and offered much expert comment and advice. Mrs. Moira Carter typed the manuscript expertly. I am indebted also to the thousands of students, many of them straight from fifth year at school, who have demonstrated that it is possible, and even enjoyable, to achieve an extensive and perceptive knowledge of organic chemistry from a short intensive course at Glasgow University taught following the pattern used in this book.

J.C.

# Preface

The topics discussed in this *Introduction to Organic Chemistry* are those areas of organic chemistry found in most GCE A-level chemistry courses. Like most books in this field it covers structure, stereochemistry, nomenclature, physical properties, reactions, and some industrial applications.

Unlike many other books the discussion of reactions emphasizes what happens to the electrons during reactions by explaining carefully and using systematically the representation of mechanisms by 'curly arrows' widely used by organic chemists for ionic reactions. Ionic reactions constitute a large part of most introductory courses in organic chemistry and they can be correlated and categorized by using a simple terminology and symbolism to help the perceptive reader learn, understand, and enjoy the subject.

The first seven chapters cover general chemical concepts relevant to organic chemistry, formal charges on atoms in differing bonding situations, structural types and naming procedures, shapes of molecules and chirality, electronegativity effects, molecules with delocalized electrons, nucleophiles, electrophiles, and representation of mechanisms. The concept of molecular orbitals, which can provide the more advanced student with a satisfying theory of bonding and reactivity, is not discussed in this introductory text.

After establishing this background of concepts and terminology, the chemistry of the different classes of organic compounds is dealt with using a functional group approach. Alkanes, alkyl halides, simple organometallics, alcohols and ethers, amines, alkenes, benzene derivatives, ketones and aldehydes, and carboxylic acids are discussed, the emphasis being on how the functional group behaves. Mechanisms are given where they are known and mechanistically related reactions are compared. Preparative routes to each class of compound can be located in the text by using the Index. After the discussion of reactions of functional groups, a review chapter brings together the small group of basic types of ionic reaction mechanism for comparison and points out some of the subtleties (such as the detailed time sequence of bond makings and breakings and the stereochemistry of reactions) which the student would meet in more advanced courses.

Throughout the text numerous problems invite the reader to think deeply about implications and extensions of facts and theories already presented, and these should be tackled as they are met. Answers are provided, but in many cases the problems provide a foretaste of a subsequent chapter. In the present state of flux on naming simple compounds it is essential that all chemists be bilingual and be able to understand both acetic acid and ethanoic acid. In many cases both names are given and used and the Index provides translations.

The book is not intended as a competitor to the compendia of data available in other texts. It is designed to reveal the coherence and system which has been detected in the reactions of organic molecules and to offer teachers and students a modern and perhaps to them novel and interesting way of looking at the science of organic chemistry.

# Contents

# List of Symbols

$\rightleftharpoons$      Indicates there is an equilibrium between these substances.

$\rightleftharpoons$      Indicates there is an equilibrium in which the right-hand substances predominate.

$\leftrightarrow$      Is placed between drawings which together attempt to represent the electron distribution in a molecule in which the electrons are delocalized.

A⸽⸽⸽⸽ B      Represents a bond between the main framework (A) and an atom (B) lying behind it.

A ◄ B      Represents a bond between the main framework (A) and an atom (B) lying in front of it.

↛      Means does not yield.

→ →      Means yields after several steps.

—$\bar{\text{N}}$—
or $\geqslant$N⏽      Represents a nitrogen atom with one lone pair of electrons and three bonding pairs.

⌒↝      Indicates how a pair of electrons has become relocated as a result of a chemical change.

$\oplus$ and $\ominus$      Indicate positive and negative charges on ions. Multiple charges as in $Cu^{2+}$ have not been circled.

# Chapter 1
## Background

### ORGANIC SUBSTANCES

Organic chemistry is that part of chemistry which deals with the compounds of carbon. You may ask why is it called organic and why are the compounds of carbon discussed separately.

The word organic nowadays has several meanings, but they all basically signify 'forming part of, or associated with something which is alive' as in the words organism, organization, organ of the body, or organic fertilizer. Early chemists studied the constituents of plants and animals and called these compounds organic compounds. They were found to contain a large proportion of carbon and indeed mostly to have frameworks of carbon atoms with hydrogen and a few other atoms added. The word organic came to be used for all carbon compounds whether isolated from living things or not, and so acquired its special meaning within chemistry.

The chemistry of carbon compounds is frequently discussed separately from the rest of chemistry for several reasons. Firstly, there are a lot of them—about as many compounds are known which contain carbon as compounds not containing carbon—and remember there are over 100 other elements. Part of the reason for the great number of known carbon compounds is that a great variety of very complex carbon-containing molecules are synthesized in living organisms and have been identified; partly it is because chemists have been interested in these natural compounds and in simpler compounds related to them and have not yet investigated the chemistry of other elements so thoroughly. However, there is a more fundamental reason which lies behind these. Carbon atoms can combine in stable groups, to form chains, rings, and other frameworks of finite size. No other single element can do this so well. Chains of carbon atoms such as those found in diamond and in hydrocarbons such as butane are very stable:

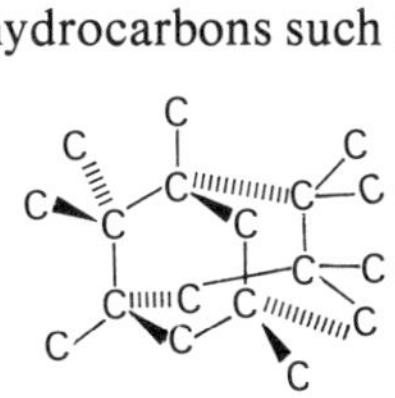

Part of the diamond lattice 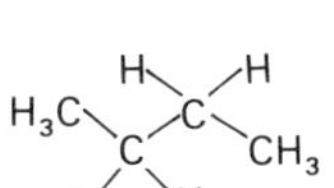      Butane

That is, they can be kept indefinitely and even when a moderate amount of energy is provided in the form of light or heat they do not decompose. Molecules with more than four silicon or boron or nitrogen atoms in a row are very rare. There are, of course, very stable crystals (the minerals in rocks and clays) containing Si, O, and Al, say, forming an infinite lattice and there are stable molecules with chains of alternating silicon and oxygen atoms (the silicone rubbers and oils), but carbon is unique in its ability to form the complete frameworks of individual molecules.

Having said this, it is important to point out that the compounds of carbon are not exceptions from the general rules of molecular behaviour which we call physical chemistry. The principles of atomic structure and bonding, of spectroscopy, of chemical energetics, and of reaction rates apply as much to carbon compounds as to those of iron or fluorine, and an understanding of these principles is essential if observations on the behaviour of carbon compounds are to become anything more than an orderless list.

In this chapter we shall review rapidly some aspects of general and physical chemistry which are essential for understanding later parts of the book. You will probably have studied, or be studying, some of these in greater detail from other books. They are reviewed briefly here to ensure that you understand the meaning and significance of some physical chemical expressions. It is important to remember that physical concepts would not exist if people had not studied molecular behaviour and that studies of molecular behaviour are made more meaningful by a knowledge of physical principles. You must not store equilibrium and activation energy in one part of your mind and nucleophiles and benzoic acid in another.

If you do not yet know what any of these things are, don't worry but read on.

## ATOMS AND COMPOUNDS

There is good, although indirect, evidence that the substance which we call sodium metal is made up of particles called atoms which are all the same. For this reason sodium is called an element. It is an example of the simplest (most elementary) kind of substance. The atoms of other elements are all different from sodium atoms. However, all atoms are made of even smaller particles—electrons, protons, and neutrons. Protons and neutrons have about the same mass; electrons have very much less. Protons are positively charged, neutrons are neutral and, electrons are negatively charged. Since atoms are neutral they always contain equal numbers of protons and electrons. All sodium atoms, for instance, contain eleven protons and eleven electrons. Most sodium atoms also contain twelve neutrons. Other kinds of sodium atoms are known, however, that contain eleven or thirteen neutrons. All three of these isotopes of sodium have eleven protons, however. It is the number of protons which identifies them as sodium atoms rather than magnesium or neon atoms, which have twelve and ten protons, respectively.

The protons and neutrons of all atoms are held tightly together in a central dense mass which we call the nucleus. The electrons occupy a much greater volume of space round the nucleus. The electrons are attracted to the nucleus by the electrostatic attraction characteristic of any two objects with opposite electric charges but they do not collapse into the nucleus. If we want to pull an electron out of the atom we must supply the assembly with enough energy to overcome the electron–nucleus attraction. There is good evidence from spectroscopy that the electrons in an atom are not all equally easy to remove. They are in groups or shells or energy levels. In the sodium atom two electrons are in the lowest energy level, eight in the second and one in the third shell. The last electron is much the easiest to remove. Electrons in the second shell are more difficult to remove and those in the first shell even more difficult.

It seems that the first shell of all atoms can accommodate only two electrons, the second only eight and there are limits to the higher shells also. The maximum capacity of all shells is an even number. This, and other evidence, suggests that electrons prefer to be in pairs. The partners in the pairs are very slightly different, so we can draw the electrons in the sodium atom using pairs of arrows with heads at opposite ends. The diagrams below indicate only the numbers of electrons in each shell.

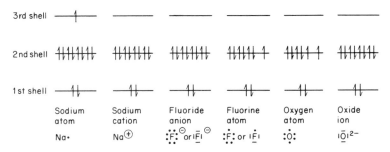

If we provide enough energy (called the ionization energy or ionization potential), we could remove an electron from the sodium atom. The easiest to remove would, of course, be that with highest energy. The remaining assembly would have a positive charge since we have removed one negative electron. We call it the sodium ion, $Na^{\oplus}$. Its ten electrons will be arranged in the shells of lowest energy as shown. Removal of a second electron from the sodium atom would be more difficult because we would have to remove it from an assembly which is already positive and also because we would have to remove an electron from a lower shell, that is, one in which the electrons have less energy than had the first one we removed.

$Na^{\oplus}$, $F^{\ominus}$ and $O^{2-}$ are well known particles which occur in crystal lattices or in solutions. We talk of ionic bonds in the compound $Na^{\oplus}F^{\ominus}$ but we would be better to talk of electrostatic attraction between each ion and the several neighbouring ions of opposite charge.

When NaF dissolves in water the ions go their separate ways but they now are surrounded by water molecules and are held on to the water molecules, again by electrostatic attraction between the charge on the ion and small charges of opposite sign in the water molecules (see Chapter 5). There is no such thing as a

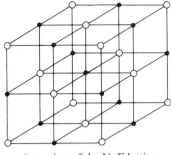

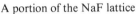

A portion of the NaF lattice

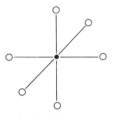

Each ion is surroun-
ded by six ions of
opposite charge

molecule of NaF in the crystal, since the word molecule means a little lump—a small structure or edifice made of several atoms, which behaves as a discrete entity.

We *can*, however, talk of a molecule of methane, $CH_4$. This group of five atoms stays together whether the methane is a solid, or a gas, or is dissolved in a solvent. Therefore, there is a force, a bond, holding the group together. We usually say that there is a bond between the carbon and each of the hydrogen atoms. This bond is much stronger than any attraction between the atoms of the methane molecule and any other extraneous atoms. This type of bond is called a covalent bond. The attraction is again fundamentally electrostatic but the attraction is between the nuclei of the bonded atoms and a number of electrons occupying a zone of space between these nuclei. A very simple diagram of the nuclei and the two electrons involved in the bond in a hydrogen molecule might look like

nucleus                               nucleus

where the density of the negative charge due to the two electrons is indicated by the density of the shading. The covalent bond is a kind of sandwich in which the nuclei are stuck together by a jam of electrons. The amount of jam in the sandwich can vary from case to case. There are molecules known with anything from one to six electrons (and in a few cases possibly eight) helping to bond together two nuclei. The most common situation and the simplest to draw is that where two electrons lie between the two nuclei. Such a bond is called a single covalent bond and is often drawn as a line, as in the hydrogen molecule, H–H. If there are more than two electrons involved, we speak of multiple bonds, for example the double bond in ethene, $H_2C=CH_2$, or the triple bond in nitrogen, $N\equiv N$. In the nitrogen molecule there are actually fourteen electrons together. Two of them are in the first shell of one nitrogen atom and two in the first shell of the other. They are not involved in the bonding. The other ten are of higher energy. Six of these are involved in the bonding and are shared, and distributed, between the two nuclei. The remaining two pairs of electrons, one pair on each nitrogen atom, do not take part in the bonding. They are called non-bonding or lone pairs.

In drawing pictures of the electron distribution in ions and molecules, the inner shell electrons are not drawn. They are seldom involved in the chemistry. All of the outermost electrons should be drawn, whether they are bonding or non-bonding. We might therefore represent nitrogen as

$$\ddot{N}\!::\!\ddot{N} \quad \text{or} \quad :N\!::\!N:$$

or, as is discussed in more detail in Chapter 2, as

$$\overline{N}\!\equiv\!\overline{N} \quad \text{or} \quad |N\!\equiv\!N|$$

where each bar represents a pair of electrons.

## ENERGY AND EQUILIBRIUM

If we allow water to freeze, its structure changes and heat is given out. If we allow carbon and oxygen to react, $C + O_2 \rightarrow CO_2$, their structures change (this time the structure of the molecules changes as well as their physical state) and again heat is given out. If we allow a battery to run down, a chemical change occurs with the production of electrical energy, and chemical energy can be turned into mechanical energy (when you tap some nitroglycerine) or into light (in a candle flame).

In all of these cases the product substances possess less energy than the reagent substances. We say they are more stable than the reagents. The relative stabilities of substances are best measured by the change in what is called the free energy, $G$, of the substance. The free energy change, $\Delta G$ (delta $G$), is numerically equal to the useful work which could be done by the conversion of all of the reagent sample into product. The change is defined as the free energy of the products minus the free energy of the reagents:

$$\Delta G = G_{\text{products}} - G_{\text{reactants}}$$

If the free energy of the products is lower than that of the reactants, that is, if the products are more stable, free energy will be transferred to the surroundings when the reaction occurs and $\Delta G$ will be negative.

Changes for which $\Delta G$ is negative tend to occur, for instance the burning of carbon or the dissolution of salt in water.

Changes for which $\Delta G$ is positive, such as $2H_2O \rightarrow 2H_2 + O_2$, will occur only if energy is acquired by the reagents from the surroundings.

If the free energies of the reagents and products are equal, $\Delta G$ will be zero. In this case there is no tendency for reagents to turn into products or products to revert to reagents. This situation can occur only if some of the product and some of the reagent are present together. A mixture of products and reagents, which has a composition such that the change in free energy for conversion of some of the one into the other is zero, is said to be at equilibrium.

Equilibrium situations are common in physics, economics, and sociology as well as chemistry but vary considerably in complexity. In each case, however, there is a rule—an equation—which is obeyed when the system is at equilibrium, as in the following examples:

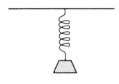

Weight of block = force exerted by spring. If the weight of the block is increased a new equilibrium position is established by extending the spring but the equation still holds

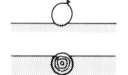

Weight of float = weight of water displaced. This is true whether the float is a balloon or a log of wood

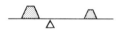

Weight × distance from fulcrum = weight × distance from fulcrum

These are elementary mechanical equilibria. We shall be more interested in equilibria involving large numbers of individuals.

Suppose on an island there are foxes and hares and the foxes eat hares and only hares. The number of foxes on the island will depend on the number of hares. If, because of a bad winter, the population of hares falls, then quickly the population of foxes will fall since there will not be enough hares for them to eat. In a good summer both populations will increase. The population of foxes is proportional to the population of hares, or more exactly the population of foxes per square mile is proportional to the population of hares per square mile, and this is true no matter what the actual populations are or how big the island is.
We can write

$$\frac{\text{population of foxes per square mile}}{\text{population of hares per square mile}} = \text{constant}$$

On another island there might be ants and hedgehogs. Again the populations would be proportional but the proportionality constant would be different because of the larger number of ants needed to feed each hedgehog.

Let us now consider a chemical equilibrium. If a solution of bromine in water is mixed with some chloroform (which does not dissolve in water), some but only some of the bromine will go into the chloroform layer if the mixture is shaken. Further shaking will not alter the distribution. If more chloroform is added, more bromine will be transferred from the water layer. An equilibrium situation is reached after shaking each new mixture. The rule in each case is

$$\frac{\text{amount of bromine per litre in the water}}{\text{amount of bromine per litre in the chloroform}} = \text{constant}$$

or

$$\frac{\text{concentration of bromine in water}}{\text{concentration of bromine in chloroform}} = \text{constant}$$

If we repeat the experiment with iodine or with petrol, a similar law would hold but the value of the constant would be different.

A more complex situation occurs when one substance can turn into another or others and also be formed from these others. For instance, if the gas $NO_2$ is compressed or cooled, pairs of molecules join up (dimerize) to give $N_2O_4$. If the pressure is released, some of the $N_2O_4$ will dissociate again. At any given pressure there will be a mixture of $NO_2$ and $N_2O_4$. In this mixture dimerizations and dissociations are taking place all the time but the composition of the mixture remains steady provided that the pressure and temperature are not changed. In this case we find that

$$\frac{\text{concentration of } N_2O_4}{square \text{ of the concentration of } NO_2} = \text{constant}$$

For the reaction $2NO_2 \rightarrow N_2O_4$ we can say that, once equilibrium is attained at a specified temperature and pressure,

$$\frac{[N_2O_4]}{[NO_2]^2} = \text{constant}$$

This constant is called the equilibrium constant and is given the symbol $K$.

Indeed, for any reaction $2A \rightarrow B$ which proceeds to give an equilibrated mixture of A and B there is an equilibrium constant defined by the equation $[B]/[A]^2 = K$. For a reaction $A \rightarrow 2B$ at equilibrium $[B]^2/[A] = K$ and for a reaction $A \rightarrow B + C$ at equilibrium $[B][C]/[A] = K$. The values and the units of the $K$s are, of course, different for each reaction.

The form of the rule for a mixture of reagents and products in equilibrium thus depends on the chemical equation for the reaction.

Let us look in more detail at the implications of an equilibrium situation. When $H_2S$ is put into water some of the molecules break up (dissociate) to give $HS^{\ominus}$ ions and $H^{\oplus}$ ions:

$$H_2S \rightleftharpoons HS^{\ominus} + H^{\oplus}$$

(The hydrogen ions will be solvated and could be represented better as hydroxonium ions, $H_3O^{\oplus}$, but the simple representation above is adequate for the present purpose.) So a mixture of all three is formed. The reaction is reversible. That is, if $HS^{\ominus}$ ions were mixed with hydrogen ions some of them would combine, but not all. A mixture of $H_2S$, $HS^{\ominus}$ and $H^{\oplus}$ would again be formed. This mixture is at equilibrium. There is no tendency for its composition to change. For such a system we find

$$\frac{[HS^{\ominus}][H^{\oplus}]}{[H_2S]} = \text{constant } (K) \approx 10^{-10} \text{ moles litre}^{-1}$$

8

The units of $K$ are not important to the present discussion but arise because the notation $[HS^\ominus]$, etc., means concentration of $HS^\ominus$ measured as moles of $HS^\ominus$ ion per litre of solution.

This equation is true of any solution containing these three substances once equilibrium has been reached, which happens rapidly in this case.

If now some $HS^\ominus$ ions were added, $[HS^\ominus]$ would momentarily increase, but very quickly some of the $HS^\ominus$ would capture protons and form $H_2S$. We can say that the reaction 'moves to the left'. This reduces the $[HS^\ominus]$ from its new value, reduces $[H^\oplus]$ and increases $[H_2S]$. Proton capture continues until again

$$\frac{[HS^\ominus][H^\oplus]}{[H_2S]} = 10^{-10}$$

although the actual values of $[HS^\ominus]$, etc. are each now different from their original values.

## RATES OF REACTION

The rate of a reaction is defined as the rate at which the concentration of one of the products or one of the reactants changes with time. Some reactions are very fast, for example $H^\oplus + OH^\ominus$ in a titration, or $CH_4 + 2O_2 \rightarrow CO_2 + 2H_2O$ in a gas flame, while some are very slow, for example C (diamonds) $+ O_2 \rightarrow CO_2$. Diamonds (fortunately) react with oxygen in the air only extremely slowly at room temperature, yet we know that if this reaction did go, the free energy of the substances would decrease ($\Delta G$ is negative) and heat would be given out. It is true in all cases that the rate of a reaction has nothing to do with the overall free energy change that would occur if the reaction proceeded to completion. $\Delta G$ values tell us nothing about reaction rates.

Diamonds do burn if they are heated in air. All reactions go faster at higher temperatures. The rates of reaction can also be affected by adding catalysts or inhibitors, but these work by changing the mechanism of the reaction so that the rate we now find is really the rate of a different process. Finally, the rates of reactions depend on the concentrations of some or all of the reagents.

In uncomplicated cases, then, we expect the rate to depend on the nature of the reactants, the concentrations of the reactants, and the temperature. We shall not be concerned again in this book with the detailed interpretation of the first two factors, but it is important here to understand why reaction rates increase with temperature.

Consider a large group of children at a party. Some will be slumped in corners. Most of them will be milling around on the ground floor where the food is. Some will be upstairs and there is a finite chance that a few will be out on the roof.

A flaskful of molecules behaves in much the same way. A few will have little energy, there is a larger fraction with medium energies, and a few with high energies.

If the temperature is increased, the total kinetic energy of the sample increases. The average energy of the molecules increases, the fraction of the population with

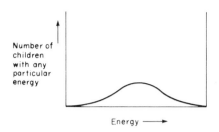

a certain relatively high energy, say $x$, increases and also the fraction of the population with energy $x$, or more than $x$, increases.

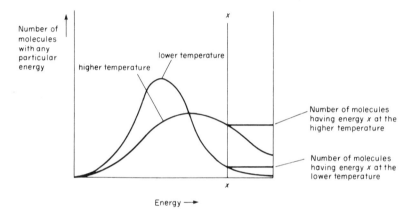

Now, what has this to do with rates of reaction? If the reaction is a simple reaction of the type A + B → C we can assume that reactions will occur only when a molecule of A collides with a molecule of B, but we can calculate how often these collisions will occur and we always find that the number of collisions per minute is greater than the number of molecules of C produced per minute. Only a small percentage of the collisions result in reaction. We must ask what the molecules in the successful collisions have that those in the unsuccessful collisions lack.

The answer to this question is energy. Only those collisions in which the reacting molecules together possess enough energy lead to reaction. Now, perhaps only a small fraction of the reagent molecules have enough or more than enough energy to react; but if the reactants were heated a larger fraction would have enough energy, so the percentage of collisions leading to reaction would increase and the reaction would go faster. It is also true that when the temperature is increased the total collision rate increases. The increase in observed reaction rate is therefore the result of (1) more collisions and (2) a greater chance that the colliding molecules possess enough energy to undergo a chemical change.

A molecule or pair of colliding molecules which has acquired enough energy to react to form products is called an activated complex. Since it is able to react we can say it is in a transition state. The difference between the average energy of the reactants and the energy possessed by molecules in the transition state is called

10

the activation energy or transition state energy. After the activated complex has been transformed into product the product molecule will still have much of this energy but will soon transfer it by collisions until it has somewhere near the average energy of the products. These collisions could be with reagents. Reaction therefore requires the temporary acquisition by one or two molecules of an amount of energy much higher than average. Reactions for which the activation energy is high will be slow since only very few of the reagent molecules will possess enough energy at any one time to react. Reactions with lower activation energies will be faster. It should be clear that the magnitude of the activation energy is in no way related to the difference in the average energies of product and reagent molecules, $\Delta G$.

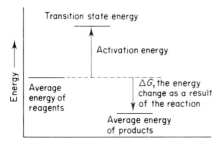

A diagram for a reaction such as $C + O_2 \rightarrow CO_2$ for which $\Delta G$ is large and negative but the rate is very slow since the activation energy is large

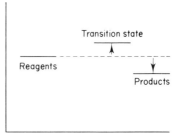

A similar diagram for a reaction such as $H^{\oplus} + OH^{\ominus} \rightarrow H_2O$ for which $\Delta G$ is small but reaction is fast since the activation energy is small

All of these ideas of bonding, free energy, equilibrium, and activation energy will be needed in later chapters.

# Chapter 2
## Electron Accounting

Almost all elements can form compounds containing carbon, but most types of organic compound can be exemplified by molecules containing elements from among the first eighteen elements of the periodic table.

*Problem 2–1. Write out these eighteen elements in order.*

Before we can discuss the structures of these compounds, we should study the types of valency and bonding that different atoms can exhibit. The simplest case is that of hydrogen. The hydrogen atom is a neutral assembly of a nucleus which has a charge of +1 (one proton) and one electron. We can draw it as H·, where the dot represents the electron, the symbol H represents the remainder (the nucleus), and there is no net charge. If, somehow, the hydrogen atom lost its one electron, the positively charged nucleus would be left, which we could write as $H^\oplus$. Throughout this book single $\oplus$ and $\ominus$ charges will be drawn with circles round them. This positively charged ion is often called *the* hydrogen ion. It is, of course, just a proton and is better called that, since there is another hydrogen ion. In solution, protons attach themselves to solvent molecules, that is they become solvated. If the solvent is water the solvated proton can be written as $H_3O^\oplus$, the hydroxonium ion. It is adequate for many purposes to write the hydrogen cation as $H^\oplus$ and to call it a proton. If the hydrogen atom were somehow to gain an electron, a negatively charged ion, $H:^\ominus$, would be formed. This ion exists, and is called the hydride ion.

Hydrogen forms compounds in which the hydrogen is in the form of a cation, a solvated proton, for example aqueous HCl, and compounds in which it is in the form of an anion, the hydride ion, for example sodium hydride, $Na^\oplus H^\ominus$. These are ionic, electrovalent, compounds. However, hydrogen also forms covalently bonded compounds, e.g. $H_2$ and gaseous HCl. In covalent compounds a pair of electrons is located in space between two nuclei, and so holds the nuclei together as an 'electrostatic sandwich'. We say that the electron pair is shared between the two atoms. For the purposes of this chapter, we shall assume that this sharing is equal, that is that each atom 'owns' one electron of the bonding pair. We draw the $H_2$ molecule as H–H, where the line represents the pair of bonding electrons.

It is convenient to use such a line to represent also the paired, but non-bonding, electrons in the hydride ion. It can be drawn, therefore, as $\bar{H}^{\ominus}$ or $H|^{\ominus}$; the position of the line in the drawing is of no significance. Non-bonding or lone pairs of electrons in all molecules will be represented as a line. If the pair is a bonding pair, the line will lead away from the atom towards some other atom. If the pair is non-bonding, the line will be peripheral to the atom:

$$\text{non-bonding electron pair } \bar{H}^{\ominus} \qquad \text{H–H bonding electron pair}$$

In nearly all compounds of hydrogen, the hydrogen has one of the following valency (outermost) electron arrangements:

$$H^{\oplus} \qquad \bar{H}^{\ominus} \qquad H\text{–}X$$

In the first case the hydrogen 'owns' one electron less than does a (neutral) hydrogen atom, so it is positive. In the second, it 'owns' one more electron than a (neutral) hydrogen atom, so it is negative. In the third case the hydrogen 'owns' a half share in a pair of electrons, i.e. it 'owns' one electron, so it is uncharged.

## VALENCY ELECTRON ARRANGEMENTS OF ELEMENTS OTHER THAN HYDROGEN

We can now apply the same approach to atoms further down the periodic table and work out the charge resulting from any electron arrangement we care to invent, but we shall restrict our discussion to valency situations which do exist. Most stable organic compounds contain atoms which possess complete or almost complete valency shells. In most compounds of boron, carbon, nitrogen, oxygen, and fluorine there are eight, or occasionally seven or six, valency electrons associated with these atoms, as in the following examples:

| BF₃ | CH₄ | NH₄⁺ | NH₃ | NO | ⁻OH |

Again, the element's letter symbol is used to represent the nucleus and all the electrons except those in the outermost shell, which are drawn using a line for a pair of electrons and a dot for any unpaired electron.

Nitrogen is a useful example to study in detail. A nitrogen atom has five outer-shell electrons. Nitrogen forms compounds in which the nitrogen may have eight, seven, or six electrons round it:

| Eight-electron arrangements | Seven-electron arrangements | Six-electron arrangement |

In the first case the nitrogen is forming four bonds and has no other electrons in the outer shell. It 'owns' a half share in four bonding pairs, that is, it 'owns' four electrons, one less than does a (neutral) nitrogen atom so the nitrogen must be positively charged. A nitrogen atom (in any of the simple compounds discussed in this book) which forms four bonds must have a positive charge. Examples are

$$
\begin{array}{c}
\text{H} \\
| \\
\text{H}-\overset{\oplus}{\text{N}}-\text{H} \\
|\\
\text{H}
\end{array}
\qquad\qquad
\overline{\text{O}}{=}\overset{\oplus}{\text{N}}{=}\overline{\text{O}}
$$

The ammonium ion, $NH_4{}^{\oplus}$    The nitronium ion, $NO_2{}^{\oplus}$

In the second case, $-\overset{|}{\text{N}}-$, the nitrogen is forming three bonds and has one lone pair of electrons. It 'owns' five electrons and so is uncharged. Examples are

$$
\begin{array}{c}
\text{H}-\overline{\text{N}}-\text{H} \\
| \\
\text{H}
\end{array}
\qquad\qquad
\text{H}-\text{C}{\equiv}\overline{\overline{\text{N}}}
$$

Ammonia      Hydrogen cyanide

In the third case, $-\overline{\overline{\text{N}}}{}^{\ominus}$, the nitrogen has two bonds and two lone pairs. It 'owns' one more electron than does a neutral nitrogen atom and so must be negative. An example is the amide ion, $^{\ominus}NH_2$, which occurs in the ionic compound sodium amide, $Na^{\oplus}{}^{\ominus}NH_2$. The charges in the seven-electron and six-electron arrangements can be deduced in the same way.

By the same argument, it is easy to build up a list of electron arrangements for all the elements and to work out the resulting charges. Some examples are given below.

$$
-\overset{|}{\text{B}}-\qquad -\overset{\oplus}{\underset{|}{\text{C}}}-\qquad -\overline{\underset{|}{\text{C}}}-\qquad -\overline{\text{N}}\qquad -\overline{\overset{\oplus}{\text{O}}}
$$

Some possible six-electron arrangements

Compounds containing atoms with only six valency electrons such as $\text{H}-\overset{\oplus}{\underset{|}{\underset{\text{H}}{\text{C}}}}-\text{H}$

or $\text{H}-\overline{\text{C}}-\text{H}$ are usually very reactive.

$$
-\overset{\bullet}{\underset{|}{\text{B}}}{}^{\ominus}\quad -\overset{\bullet}{\underset{|}{\text{C}}}-\quad -\overset{\bullet}{\underset{|}{\text{N}}}{}^{\oplus}\quad -\overset{\bullet}{\underset{|}{\text{N}}}-\quad -\overset{\bullet}{\overline{\text{O}}}{}^{\oplus}\quad -\overline{\overset{\bullet}{\text{O}}}\bullet\quad |\overline{\overline{\text{F}}}\bullet
$$

Some seven-electron arrangements

Atoms or groups of atoms with an odd number of valency electrons are called radicals. Thus the hydrogen atom, $\text{H}\bullet$, the methyl radical, $\text{H}-\overset{\bullet}{\underset{\text{H}}{\text{C}}}-\text{H}$, and the nitrogen oxide molecule, $\overset{\bullet}{\text{N}}{=}\overline{\text{O}}$, are all radicals. Radicals are usually rather reactive.

$$
\overset{\diagup}{\underset{|}{\text{P}}}\diagdown \qquad \overset{\diagup}{\underset{|}{\text{S}}}\diagdown \qquad \overline{\underset{|}{\text{S}}}\diagdown
$$

Some ten- and twelve-electron arrangements

Only elements in the second or lower rows of the periodic table can form compounds in which more than eight valency electrons are associated with any atom.

By far the most important are the eight electron arrangements tabulated below.

$$-\overset{|}{\underset{|}{B}}{}^{\ominus} \qquad -\overset{|}{\underset{|}{C}}- \qquad -\overset{|}{\underset{|}{N}}{}^{\oplus}$$

$$-\overline{\underset{|}{C}}{}^{\ominus} \qquad -\overline{\underset{|}{N}}- \qquad -\overline{\underset{|}{O}}{}^{\oplus}$$

$$-\overline{\overline{N}}{}^{\ominus} \qquad -\overline{O}- \qquad -\overline{\underset{|}{F}}{}^{\oplus}$$

$$-\overline{\underset{|}{O}|}{}^{\ominus} \qquad -\overline{F}|$$

$$|\overline{O}|^2 \qquad |\overline{F}|^{\ominus}$$

Check that the charges tally with the electron arrangement round each atom. Examples of assemblies of atoms from this table include neutral molecules, such as

$$\begin{array}{c} H \\ | \\ H-\overset{|}{C}-H \\ | \\ H \end{array} \qquad \begin{array}{c} \overline{\phantom{N}} \\ H-\overset{|}{N}-H \\ | \\ H \end{array} \qquad H-\overline{\overline{O}}-H \qquad H-\overline{F}|$$

   Methane        Ammonia         Water         Hydrogen
                                                fluoride

cations such as

$$\begin{array}{c} H \\ | \\ H-\overset{\oplus}{N}-H \\ | \\ H \end{array} \qquad \begin{array}{c} \overline{\phantom{O}}{}^{\oplus} \\ H-\overset{|}{O}-H \\ | \\ H \end{array}$$

   Ammonium ion      Hydroxonium ion

and anions such as

$$H-\overline{\underline{O}}|{}^{\ominus} \qquad H-\overline{\overline{N}}{}^{\ominus}H \qquad \begin{array}{c} \overline{\phantom{C}}{}^{\ominus} \\ H-\overset{|}{C}-H \\ | \\ H \end{array} \qquad \begin{array}{c} H \\ | \\ H-\overset{\ominus}{B}-H \\ | \\ H \end{array}$$

You will have met many of these substances before but may not know their names.

Positive ions containing the units

$$-\overset{|}{\underset{|}{C}}{}^{\oplus} \qquad -\overset{|}{\underset{|}{N}}{}^{\oplus} \qquad -\overline{O}{}^{\oplus} \qquad -\overline{Br}{}^{\oplus}$$

are called carbonium ions, ammonium ions, oxonium ions, and bromonium ions, respectively. All ions whose names end in onium, enium, inium or ium are cations. Note that carbonium ions are different from the others in having only six electrons altogether round the carbon and some chemists therefore prefer to call them carbenium ions.

*Problem 2–2. Draw the electron arrangements typical of phosphonium ions, sulphonium ions, and chloronium ions.*

The names of anions end in ate, ite, ide, etc.; for example: chloride, $|\overline{\underline{Cl}}|^{\ominus}$; hydroxide, $H-\overline{\underline{O}}|^{\ominus}$; amide, $^{\ominus}\overline{N}H_2$; methide, $^{\ominus}\overline{C}H_3$; and tetrahydroaluminate, $^{\ominus}AlH_4$. Ions such as $^{\ominus}\overline{C}H_3$ and $^{\ominus}\overline{C}{\equiv}N$, which have a negative charge on the carbon, are called *carbanions*.

*Problem 2–3. Draw pictures like those above for the hydrosulphide ion and the tetrafluoroborate ion, working out the structures from the names.*

## ISOELECTRONIC MOLECULES

The substances $^{\ominus}BH_4$ (tetrahydroborate ion), $CH_4$ (methane), and $^{\oplus}NH_4$ (ammonium ion) have the structures

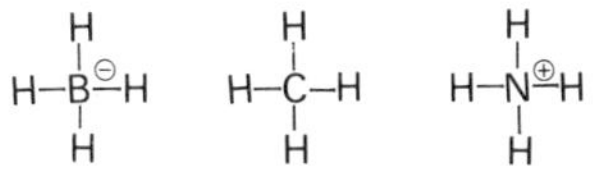

$$\begin{array}{ccc} H & H & H \\ | & | & | \\ H{-}\overset{\ominus}{B}{-}H & H{-}C{-}H & H{-}\overset{\oplus}{N}{-}H \\ | & | & | \\ H & H & H \end{array}$$

They have identical electron arrangements. They are said to be *isoelectronic*. In fact, the only difference between them is the number of protons in the nuclei of the central atoms. Carbon has one proton more than boron and nitrogen one more again. Is that in accord with the charges on these molecules? Note that the boron atom of the tetrahydroborate ion is negative, even though it has no lone pairs.

*Problem 2–4. What molecules or ions are isoelectronic with the tetrahydro-aluminate ion, $^{\ominus}AlH_4$, and with the nitronium ion, $^{\oplus}NO_2$?*

Check that the electron arrangements shown below imply the charges indicated:

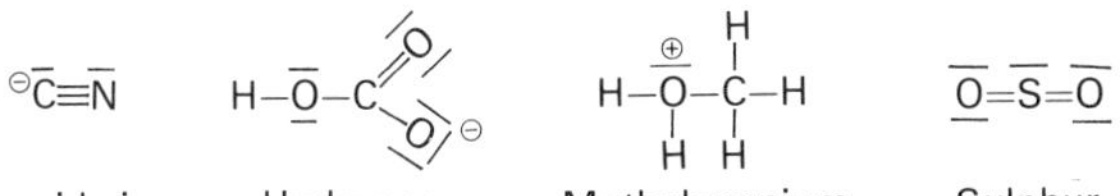

| $^{\ominus}\overline{C}{\equiv}\overline{N}$ | $H{-}\overline{\underline{O}}{-}C\overset{\diagup O}{\underset{\diagdown \underline{O}|^{\ominus}}{}}$ | $H{-}\overset{\overset{\oplus}{H}}{\underset{\underset{H}{|}}{\overline{O}}}{-}\overset{|}{\underset{|}{C}}{-}H$ (H H) | $\overline{\underline{O}}{=}S{=}\overline{\underline{O}}$ |
|:---:|:---:|:---:|:---:|
| Cyanide ion | Hydrogen-carbonate ion | Methyloxonium ion | Sulphur dioxide |

For example, in the methyloxonium ion:
  all of the hydrogen atoms own a half share of a bond (= 1 electron each), therefore they are uncharged since a hydrogen atom has 1 electron;
  the carbon atom owns a half share in four bonds (= 4 electrons), therefore it is uncharged since a carbon atom has 4 electrons;
  the oxygen atom owns one lone pair plus a half share in 3 bonds (= 5 electrons), therefore it is positive since an oxygen atom has 6 electrons.

16

Molecules isoelectronic with some of these listed above include the following:

$$N_2 \quad C_2^{2-} \quad CO \quad HNO_3 \quad HO-\overset{\overset{\displaystyle O}{\|}}{C}-F \quad N_2H_5^{\oplus} \quad ClO_2^{\oplus} \quad CH_3NH_2$$

*Problem 2–5. Draw out full electron arrangements for each of the molecules in this list as was done above.*

## COUNTING UP ELECTRONS

You may be wondering how these pictures were arrived at. Consider the nitrite ion, which is known to have the molecular formula $NO_2^{\ominus}$ and to have a structure in which the oxygen atoms are bonded to the nitrogen but not to one another. The ion can be imagined as an assembly of a nitrogen atom, two oxygen atoms and one more electron:

| | |
|---|---|
| nitrogen atom | 5 |
| 2 oxygen atoms | $2 \times 6$ |
| $\ominus$ charge | 1 |
| | $\overline{18}$ valency electrons altogether |

If we assume that the eighteen electrons form nine pairs, bonding or non-bonding, we can invent a possible distribution of the nine pairs as

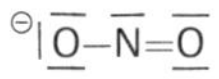

$$^{\ominus}|\overline{\underline{O}}-\overline{N}=\overline{\underline{O}}$$

This arrangement would give each atom a complete octet and would imply that one oxygen atom was negative and the other atoms neutral. It is not the only arrangement possible and it may not be the true arrangement, but it is a reasonable one.

*Problem 2–6. Work out the total number of valency electrons and suggest probable electron arrangements and then charges on each atom for the ions $NO^{\oplus}$, $[OCN]^{\ominus}$, $[CNO]^{\ominus}$, $N_3{}^{\ominus}$, and for the nitrate ion, $NO_3{}^{\ominus}$. In the ions $[OCN]^{\ominus}$, $[CNO]^{\ominus}$ and $N_3{}^{\ominus}$, the atoms are in a row. In the nitrate ion the oxygen atoms are each attached to the nitrogen but not to one another. You may find that there are more charges than you might expect, or that more than one arrangement can be drawn. This problem will be discussed later (Chapter 6).*

The pictures for electron arrangements drawn above employ a very simple and general symbolism of lines, dots, and charge signs. The pictures are quite independent of how, and even if, the compounds can be made. Some books introduce the concept of a dative or coordinate bond, but it is not needed in organic chemistry and can be confusing.

In our accounting of electrons we have seen that the charge on an atom in a compound is a result of the electron arrangement round a particular nucleus.

Changing the charge in an isoelectronic series of compounds by changing the nucleus has no effect on the structure and bond angles of the molecule (which depend mainly on the electronic arrangement), but has a profound effect on the chemical reactivity (compare $CH_4$ and $NH_4^{\oplus}$).

*Problem 2–7. Which of the following is isoelectronic with ammonia: $CH_3^{(\cdot)}$, $BH_3$, $PH_3$, $H_3O^{\oplus}$?*

*Problem 2–8. Draw in the charges implied by the electron arrangement in the following ions and molecules:*

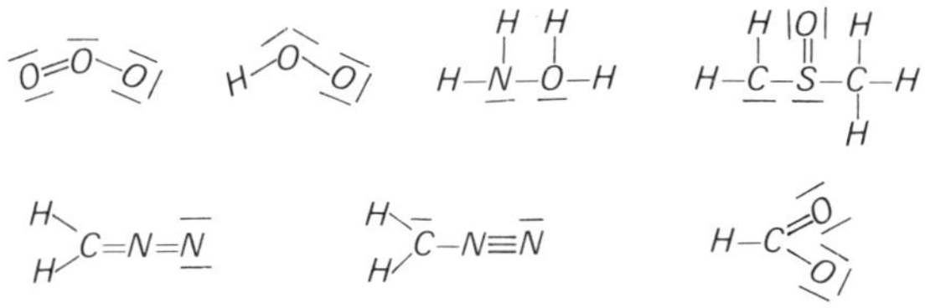

*Problem 2–9. Work out the total number of valency electrons in the ion $[O_2H_3]^{\oplus}$ and suggest a structure for this ion showing all of the valency electrons.*

*Problem 2–10. Work out the total number of valency electrons in the molecule $NO_2$ and suggest a structure for this molecule (showing all the valency electrons) in which the nitrogen lies between the oxygens and the molecule is not cyclic. Is the molecule a radical? Do the same for the non-cyclic molecule $N_2O$. The oxygen is at the end. Is this molecule a radical?*

## ANSWERS TO PROBLEMS IN CHAPTER 2

*Problem 2–1.* The first eighteen elements in the periodic table are as follows:

|     |     |     |     |     |     |     |     |
| --- | --- | --- | --- | --- | --- | --- | --- |
| H   |     |     |     |     |     |     | He  |
| Li  | Be  | B   | C   | N   | O   | F   | Ne  |
| Na  | Mg  | Al  | Si  | P   | S   | Cl  | Ar  |

*Problem 2–2.* Phosphonium, sulphonium and chloronium ions can be represented in a general way by, respectively

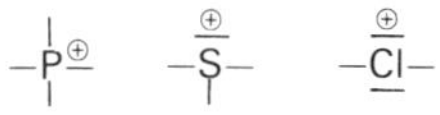

*Problem 2–3.* Hydrosulphide ion, $H-\overline{\underline{S}}|^{\ominus}$. Tetrafluoroborate ion,

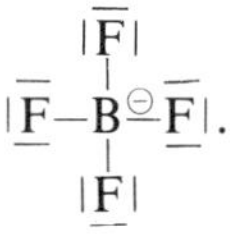

*Problem 2–4.* $^{\ominus}AlH_4$    $SiH_4$    $^{\oplus}PH_4$.    $\overline{\underline{O}}=B=\overline{\underline{O}}^{\ominus}$    $\overline{\underline{O}}=C=\overline{\underline{O}}$    $\overline{\underline{O}}=\overset{\oplus}{N}=\overline{\underline{O}}$.

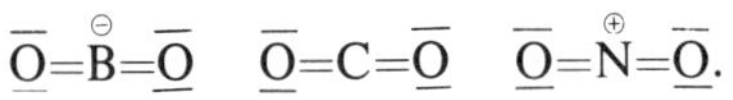

18

*Problem 2–5.* $\overline{N}\equiv N$ $^{\ominus}\overline{C}\equiv\overline{C}^{\ominus}$ $^{\ominus}\overline{C}\equiv\overline{O}^{\oplus}$ isoelectronic with $^{\ominus}\overline{C}\equiv\overline{N}$.

$H-\overline{O}-\overset{\oplus}{N}\overset{\displaystyle\nearrow O}{\underset{\displaystyle\searrow O}{}}{}_{\ominus}$ $H-\overline{O}-C\overset{\displaystyle\nearrow O}{\underset{\displaystyle\searrow \overline{F}}{}}$ isoelectronic with $H-\overline{O}-C\overset{\displaystyle\nearrow O}{\underset{\displaystyle\searrow \overline{O}}{}}{}_{\ominus}$.

$H-\overline{N}-\overset{H}{\underset{H}{\overset{|}{\underset{|}{N}}}}\overset{\oplus}{}H$   $H-\overline{N}-\overset{H}{\underset{H}{\overset{|}{\underset{|}{C}}}}-H$ isoelectronic with $H-\overset{\oplus}{O}-\overset{H}{\underset{H}{\overset{|}{\underset{|}{C}}}}-H$.

$\overline{O}=\overset{\oplus}{\overline{Cl}}=\overline{O}$ isoelectronic with $\overline{O}=S=\overline{O}$.

*Problem 2–6.*

$NO^{\oplus}$  $5+6-1=10$ valency electrons  $\overline{N}\equiv\overline{O}^{+}$

$|OCN|^{\ominus}$  $6+4+5+1=16$ valency electrons  $\overline{O}=C=\overline{N}^{\ominus}$ or $^{\ominus}|\overline{O}-C\equiv N$

$[CNO]^{\ominus}$  $4+5+6+1=16$ valency electrons  $^{\ominus}\overline{C}\equiv\overset{\oplus}{N}-\overline{O}|^{\ominus}$

$N_3{}^{\ominus}$  $3\times5+1=16$ valency electrons  $^{\ominus}\overline{N}=\overset{\oplus}{N}=\overline{N}$ or $^{\ominus}\overline{N}=\overset{\oplus}{N}=\overline{N}^{\ominus}$

or $\overline{N}\equiv\overset{\oplus}{N}-\overline{N}|^{2-}$

$NO_3{}^{\ominus}$  $5+3\times6+1=24$ valency electrons  (structures)

*Problem 2–7.* Only $H-\overset{\oplus}{\underset{H}{\overset{|}{O}}}-H$ has exactly the same electron arrangement as ammonia. $H-\overset{\underset{H}{|}}{\overline{P}}-H$ looks similar but since P is below N in the Periodic Table there is an extra shell of eight electrons in the phosphorus compound.

*Problem 2–8.* (structures) $H-\overset{H}{\underset{H}{N}}-\overset{\oplus}{O}-H$   $H-\overset{H}{\underset{}{C}}-\overset{\ominus}{S}-CH_3$

(structures)

*Problem 2–9.* $O_2H_3{}^{\oplus}$ has $2\times6+3-1=14$ valency electrons. $H-\overline{O}-\overset{\underset{H}{|}}{\overset{\oplus}{O}}-H$.

*Problem 2–10.* $NO_2$ has $5+2\times6=17$ valency electrons. Since this is an odd number, this molecule is a radical. It is, however, stable at room temperature.

$$\overline{O}=\overset{\bullet}{N}=\overline{O}$$

$N_2O$ has $2\times5+6=16$ valency electrons. It is not a radical.

$$^{\ominus}\overline{N}=\overset{\oplus}{N}=\overline{O} \quad \text{or} \quad \overline{N}\equiv\overset{\oplus}{N}-\overline{O}|^{\ominus}$$

# Chapter 3

# Classifying and Naming Compounds

Now that the basic electronic arrangements round atoms and the resulting charges are clear, we can explore the types of ions and molecules which would be formed by joining together on paper various atoms in their various valency states. If we consider compounds containing carbon, hydrogen, and the other elements in the first row of the Periodic Table and if we remember that double and triple bonds are possible as in $H_2C=\bar{O}$ or $H-C\equiv\bar{N}$, we can extend the range of molecules and ions discussed in Chapter 2 considerably. Even if we consider only assemblies of hydrogen atoms, one or two carbon atoms and one or two atoms of other first-row elements we can invent a long list of structures such as

$$
\begin{array}{c}
\underset{\underset{H}{|}}{\overset{\overset{H}{|}}{H-C-N-H}} \overset{H}{\underset{H}{\oplus}}
\end{array}
\qquad
\underset{\underset{H}{|}}{\overset{\overset{H}{|}}{H-C\cdot\cdot\bar{N}=\bar{O}}}
\qquad
\underset{H}{\overset{H}{\diagdown}}C=\bar{N}\diagdown_{\underline{O}-H}^{-}
\qquad
\underset{H}{\overset{H}{\diagdown}}\overset{\oplus}{C}-\bar{O}\diagdown_{H}
\qquad
CH_3-CH_2-\overset{\ominus}{\underline{\bar{O}}\textrm{i}}
$$

*Problem 3–1. Extend this list as far as you can. You should be able to construct over thirty structures using one or two carbon atoms, one or two other atoms from the first ten in the Periodic Table, and as many hydrogen atoms as you need. The molecules whose structures you draw may not all be isolatable.*

Obviously, if we used more carbon atoms, the number and variations would increase rapidly. However, one of the important facts of organic chemistry is that C–H and C–C bonds do not usually break easily, so the reactive parts of molecules are those where there are C=C or C≡C bonds or other elements. If, therefore, we consider a large molecule such as

$$
\begin{array}{c}
CH_3-CH_2-CH-CH_3 \\
| \\
CH_2-O-H
\end{array}
$$

the parts of the molecule made of C–H and C–C bonds will be relatively inert and the important reactions of this molecule will be reactions due to the presence of the −O−H group and these reactions will also be exhibited by simpler molecules such as $CH_3OH$ or $CH_3CH_2OH$.

Any molecule possessing a given reactive group or *functional group* of atoms (such as the —O—H group in the examples above) will behave in a typical way, no matter what the rest of the molecule may be. There may be complications if the molecule is very large or if there are two or more functional groups near one another in the molecule (phenols, p. 129, acids, p. 161). We can, however, study the properties of small molecules containing a functional group and feel safe in assuming that a larger molecule with the same group will behave in much the same way. This means that it is convenient to classify compounds according to their functional groups and then to give names to these classes—usually names related to the name of the group.

## NAMES OF COMPOUND TYPES

Compounds containing carbon and hydrogen only are called *hydrocarbons*. If there are no double or triple bonds they are called *alkanes* or *saturated hydrocarbons*. The word *saturated* is used to describe any molecule which has no double or triple bonds.

Hydrocarbons with one double bond are called *alkenes*. There is no simple name for the group $\diagup C = C \diagdown$.

Hydrocarbons with one triple bond are called *alkynes*. There is no simple name for the group $-C \equiv C-$.

$$CH_3CH_2CH_3 \qquad CH_3CH=CH_2 \qquad CH_3C \equiv CH$$
$$\text{An alkane} \qquad\qquad \text{An alkene} \qquad\qquad \text{An alkyne}$$

Other molecules consisting only of C and H atoms include *carbenes*, which contain the group $-\overset{..}{C}-$, *carbonium ions*, which contain the group $-\overset{\oplus}{\underset{|}{C}}-$, *carbanions*, which contain the group $-\overset{\ominus}{\underset{|}{C}}$, and alkyl radicals, which contain the group $-\overset{.}{\underset{|}{C}}-$. Thus the ethyl cation is a carbonium ion and the ethyl anion, or ethide ion, is a carbanion.

$$CH_3-\overset{..}{C}-CH_3 \qquad \overset{\oplus}{CH_3CH_2} \qquad \overset{\ominus}{CH_3CH_2} \qquad \overset{.}{CH_3CH_2}$$
$$\text{A carbene} \qquad \text{A carbonium} \qquad \text{A carbanion} \qquad \text{An alkyl radical}$$
$$\text{ion} \qquad\quad \text{(the ethide ion)} \qquad \text{(the ethyl radical)}$$
$$\text{(the ethyl cation)}$$

If, in our imagination, we delete one hydrogen nucleus from an alkane, we obtain a group of atoms such as

$$CH_3- \qquad \text{or} \qquad CH_3-\overset{\displaystyle H}{\underset{\displaystyle CH_3}{\overset{|}{\underset{|}{C}}}}-$$

Such groups are called *alkyl groups*; the symbol R is often used to represent any alkyl group.

If, in our imagination, we replace one or more of the hydrogen atoms of an alkane by halogen atoms, we obtain *halogenoalkanes* such as difluoromethane,

$CH_2F_2$. If there is only one halogen atom, the molecule is often called an *alkyl halide*, for example, $CH_3Br$ may be called methyl bromide or bromomethane. Such molecules contain a *halogeno group*, e.g. Cl–.

The H–O– group is called a *hydroxyl group*. Most molecules in which this is attached to carbon are called *alcohols*, e.g. $CH_3OH$, or in general ROH.

Molecules containing the group –O– attached to *two* carbon atoms are called *ethers*, e.g. $CH_3$–O–$CH_3$, or in general R–O–R′.

Molecules containing the group –O$^\ominus$ are called *alkoxide ions*, e.g. $CH_3$–O$^\ominus$ or in general R–O$^\ominus$.

Molecules containing the group –$\overset{\oplus}{\underset{|}{O}}$– are *oxonium ions*, e.g. $CH_3$–$\overset{\oplus}{\underset{\underset{CH_3}{|}}{O}}$–$CH_3$.

The chemistry of all of these substances is usually discussed together with that of alcohols and ethers.

Molecules containing an *amino group*, –$NH_2$, –$\overset{H}{\underset{|}{N}}$–, or –$\overset{|}{\underset{|}{N}}$–, attached to carbon are called *amines*, e.g. $CH_3NHCH_3$.

Molecules containing the group –$\overset{|\oplus}{\underset{|}{N}}$– are called *ammonium ions*, e.g. $(CH_3)_2\overset{\oplus}{N}H_2$.

The group $\text{>C=O}$ is called a *carbonyl group*. If it is attached to two carbon atoms the compound containing it is called a *ketone*, e.g. $CH_3-\overset{\overset{O}{\|}}{C}-CH_3$ or in general R–$\overset{\overset{O}{\|}}{C}$–R′. If it is attached to one or two hydrogen atoms the compound containing it is called an *aldehyde*, e.g. H–$\overset{\overset{O}{\|}}{C}$–H or $CH_3CH_2-\overset{\overset{O}{\|}}{C}-H$ or in general R–$\overset{\overset{O}{\|}}{C}$–H.

Some compounds have two groups close together which affect one another so much that these compounds have been given names of their own. Most important are the following:

*carboxylic acids*, containing a *carboxyl group* $-C\overset{\nearrow O}{\underset{\searrow O-H}{}}$ made up of a carbonyl group and a hydroxyl group;

*esters*, containing an *ester group* $-C\overset{\nearrow O}{\underset{\searrow O-R}{}}$ made up of a carbonyl group and an ether group;

*amides*, containing an *amide group* $-C\overset{\nearrow O}{\underset{\searrow N<}{}}$ made up of a carbonyl group and an amino group;

and *carboxylate ions*, containing a *carboxylate group* $-C\overset{\nearrow O}{\underset{\searrow O^\ominus}{}}$.

These names for functional groups and for compound classes are listed in the table below. Note that, in the examples above and in the table, the lone pairs and some of the bonding pairs of electrons have not been drawn in. This is done here and later for convenience. Other contractions are discussed on p. 27. There is no harm in drawing in all of the electrons as has been done in the first column of the table.

| Group | Name of group | Name of compound class | Example |
|---|---|---|---|
| None | — | Alkane | $CH_3CH_2CH_3$ |
| $>C=C<$ | None | Alkene | $CH_3-CH=CH_2$ |
| $-C\equiv C-$ | None | Alkyne | $CH_3-C\equiv CH$ |
| $-\underline{Br}|$ | Bromo group | Bromoalkane or alkyl bromide | $CH_3-Br$ |
| $-\underline{O}H$ | Hydroxyl group | Alcohol | $CH_3CH_2-O-H$ |
| $-\underline{O}-$ | Ether group | Ether | $CH_3CH_2-O-CH_3$ |
| $-\overline{N}H_2$ $-\overline{N}H-$ $-\overline{N}<$ | Amino group | Amine | $CH_3-NH_2$ |
| $|\underset{\|}{\overset{|\underline{O}|}{C}}-$ | Carbonyl group | Ketone | $CH_3-\overset{O}{\overset{\|}{C}}-CH_2CH_3$ |
|  |  | Aldehyde | $CH_3-\overset{O}{\overset{\|}{C}}-H$ |
| $-C\overset{\nearrow O}{\underset{\searrow O-H}{}}$ | Carboxyl group | Carboxylic acid | $CH_3-C\overset{\nearrow O}{\underset{\searrow O-H}{}}$ |
| $-C\overset{\nearrow O}{\underset{\searrow O-R}{}}$ | Ester group | Ester | $CH_3-C\overset{\nearrow O}{\underset{\searrow O-CH_3}{}}$ |
| $-C\overset{\nearrow O}{\underset{\searrow \overline{N}R_2}{}}$ | Amide group | Amide | $CH_3-\overset{O}{\overset{\|}{C}}-N\overset{CH_3}{\underset{H}{}}$ |
| $-N\overset{\nearrow O}{\underset{\searrow O^\ominus}{}}$ | Nitro group | Nitroalkanes | $CH_3-NO_2$ |

The classification of a compound depends on the functional groups and not on the carbon framework. Inspect the following and confirm their classification:

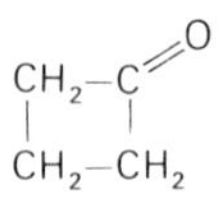

$$\underset{CH_2-CH_2}{CH_2-C\overset{\diagup O}{}}$$

A ketone

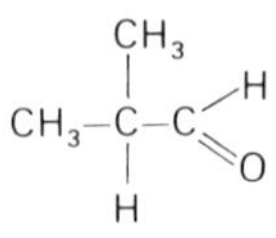

$$CH_3-\underset{H}{\overset{CH_3}{C}}-C\overset{\diagup H}{\underset{\searrow O}{}}$$

An aldehyde

$$CH_3\underset{}{\overset{CH_3}{CH}}-CH_2-NH_2$$

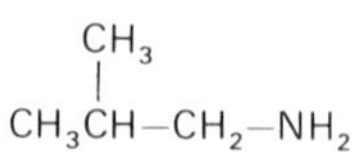

An amine

$$CH_3-CH_2-O-\underset{\underset{O}{\|}}{C}-CH_3 \qquad CH_3-CH_2-O-CH_3$$

An ester · · · · · · · · An ether · · · · · · · · An alcohol

It is important to be able to recognize functional groups however they are drawn. The ester above might have been drawn as

$$CH_3-CH_2-O-\underset{\underset{CH_3}{|}}{C}=O \quad \text{or} \quad CH_3-C\overset{\diagup O}{\diagdown O-CH_2CH_3} \quad \text{or} \quad CH_3-C=O \\ | \\ CH_2-O \\ | \\ CH_3$$

*Problem 3–2. Identify the functional group and hence the compound class for the following examples:*

$$CH_3-CH-CH_3 \qquad F-CH_3 \qquad H-C-CH_2 \qquad CH_2-CH_2 \qquad CH_3 \\ | \qquad\qquad\qquad\qquad \| \quad | \qquad\quad | \quad | \qquad\qquad | \\ O-H \qquad\qquad\qquad\qquad H-C-CH_2 \qquad CH_2-O \qquad CH_3-N-H$$

$$H-C-CH_3 \quad H-O-C-CH_3 \quad CH_3-O-C-CH_3 \qquad H\ O \qquad \overset{\oplus}{O}=N-CH_2CH_3 \\ \| \qquad\qquad \| \qquad\qquad\qquad \| \qquad\quad |\ \| \qquad\qquad\quad | \\ O \qquad\qquad O \qquad\qquad\qquad O \qquad H-N-C-CH_3 \qquad O^{\ominus}$$

$$H-O-CH_3 \qquad\qquad (CH_3)_3N \qquad\qquad H-C{\equiv}C-H$$

The functional groups may be attached to, or built into, carbon frameworks of various types:

$$CH_3CH_2CH_2-CH-CH_3 \qquad CH_2-CH_2 \\ | \qquad\qquad\quad | \qquad\quad\diagdown \\ OH \qquad\qquad CH_2-CH_2\diagup CHOH \qquad CH_3-\underset{\underset{CH_3}{|}}{\overset{\overset{CH_3}{|}}{C}}-CH_2OH$$

The above molecules are all alcohols but in the first case the carbon skeleton of the molecule is a row of carbon atoms, in the second case a ring and in the third a row with branching carbon groups. We describe the skeletons as acyclic, cyclic, branched acyclic, etc.

### NAMES FOR INDIVIDUAL COMPOUNDS

Having discussed the names of functional groups and types of skeletons we are in a position to devise names for individual compounds. Compounds often have trivial (common) names such as acetone or menthol, often given to them before their structures were known. There are advantages in naming structures in a systematic way which can be extended to the more complex cases, and this requires the invention and use of a structure-based code. Such a code and rules for applying it have been agreed internationally. The main rules are that the name is based on a code word for the longest carbon chain or the basic ring system in the molecule with prefixed or suffixed code words to indicate groups attached to it and numbers to indicate where on the chain these groups are attached.

The code words for chains of increasing length are as follows:

| $C_1$ | $C_2$ | $C_3$ | $C_4$ | $C_5$ | $C_6$ | $C_7$ | $C_8$ |
|-------|-------|-------|-------|-------|-------|-------|-------|
| meth- | eth- | prop- | but- | pent- | hex- | hept- | oct- ...etc. |

The latter ones are derived from Greek numbers.

If the skeleton is cyclic, the prefix cyclo- is put in; if the molecule is a saturated hydrocarbon, the suffix -ane is added:

$$CH_3-CH_3$$

Ethane

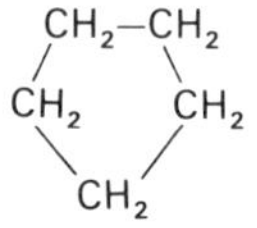

Cyclopentane

If there is a C=C double bond, the suffix is -ene; if there are two C=C double bonds it is -diene. If there is a C≡C triple bond, the suffix is -yne. The position of the multiple bond is indicated by numbering the carbon atoms of the chain and inserting, before the code for the attached group, the number of the carbon atom to which it is attached, or the number of the first carbon atom in the case of double or triple bonds:

$$CH_3-CH=CH-CH_3$$

But-2-ene

$$CH_2=CH-CH=CH_2$$

Buta-1,3-diene

$$CH_3-CH_2-C≡CH$$

But-1-yne

The compound $C_6H_6$, which consists of a ring of six CH groups as shown on the left below:

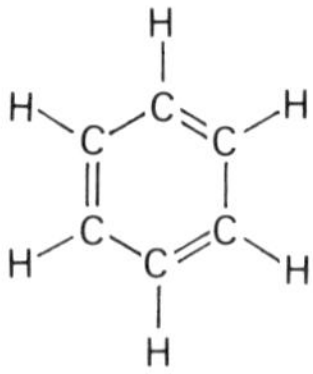

Benzene

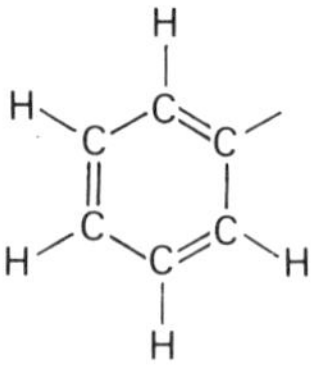

Phenyl group

has properties rather different from those of simple alkenes. It is given the name *benzene*. The group $C_6H_5-$ related to it is called the *phenyl group*.

The names of alkyl groups attached to the main chain or ring are derived from the code for the appropriate carbon chain length plus the syllable -yl. In compound names they appear as prefixes as in the following examples:

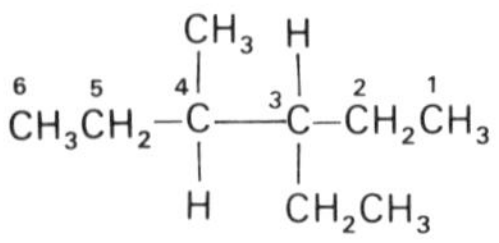

3-Ethyl-4-methylhexane

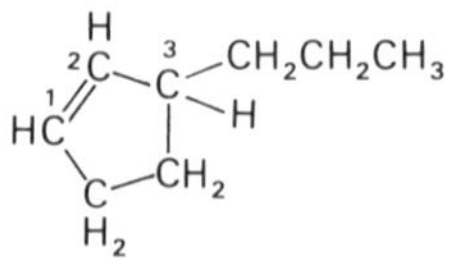

3-Propylcyclopent-1-ene

If the attached groups are halogens, their names appear in prefixes. If the attached group is an amino group, the code amino- appears as a prefix. If the attached group is a hydroxyl group, either the prefix hydroxy- or the suffix -ol is added. If a ketonic carbonyl group is present, the suffix -one is added. If an aldehyde carbonyl is present, the suffix -al is added. If a carboxyl group is present, the name ends in -oic acid. All of these suffixes come after the appropriate -an-, -en-, or -yn- suffix.

The prefix and suffix codes are summarized in the following table.

| Group | Name of group | Prefix code | Suffix code |
|---|---|---|---|
| None | — | — | -ane |
| $>C=C<$ | None | — | -ene |
| $-C\equiv C-$ | None | — | -yne |
| $-\bar{Br}|$ | Bromo group | Bromo- | -yl bromide |
| $-\bar{O}H$ | Hydroxyl group | Hydroxy- | -ol |
| $-\bar{O}-$ | Ether group | —* | — |
| $-\bar{N}H_2$ $-\bar{N}H-$ $-\bar{N}<$ | Amino group | Amino- | -ylamine |
| $-\overset{|O|}{\underset{}{C}}-$ | Carbonyl group | Oxo- | -one |
| | | Oxo- | -al |
| 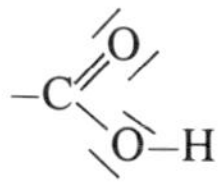 $-C\overset{O}{\underset{O-H}{}}$ | Carboxyl group | — | -oic acid |
| $-C\overset{O}{\underset{O-R}{}}$ | Ester group | —† | — |
| 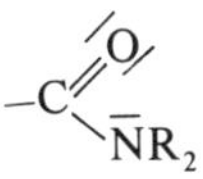 $-C\overset{O}{\underset{NR_2}{}}$ | Amide group | — | -amide |
| 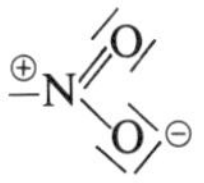 $\overset{\oplus}{N}\overset{O}{\underset{O^{\ominus}}{}}$ | Nitro group | Nitro- | — |

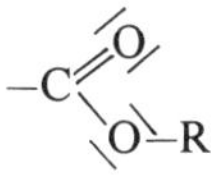

* For examples of naming of ethers see pp. 96, 101 and 126.
† For examples of naming of esters see p. 163.

There are rules for arranging the parts of the name like those for arranging ambassadors at a reception and there are alternatives, short cuts and local variations, some of which will be discussed later. Here we shall unravel some code names and draw appropriate pictures.

Consider the compound called 3-amino-4-chlorobutanal. The formula must have an amino group attached to carbon number 3 (C-3) and a chloro group attached to C-4 of a four-carbon chain with no double or triple bonds and with an aldehyde group at C-1. Compare the names and the formulae of the other examples below.

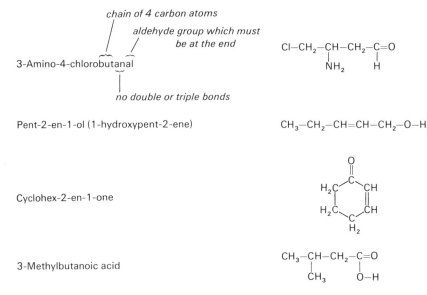

The longest chain of carbon atoms is a chain of five ...                     -pent-
There are no C=C or C≡C groups .............. .                              -pentan-
The compound is a ketone .................... .                              -pentanone
The carbonyl group is at C-2 or at C-4, depending
which end of the chain we number from; the rules
require numbering from the end which gives the smaller
numbers.................................                                     -pentan-2-one
There is a methyl group attached at C-3 and a chloro
group at C-1; these can go in as prefixes; prefixes are
put in alphabetical order ......................1-chloro-3-methylpentan-
2-one

*Problem 3–3. Devise names for*

$$CH_3CH_2\overset{\overset{\displaystyle CH_3}{|}}{C}=CH_2 \qquad HO-\overset{\overset{\displaystyle O}{\|}}{C}-\overset{\overset{\displaystyle CH_3}{|}}{CH}-CH_2-NH_2 \qquad HO-CH_2-\overset{\overset{\displaystyle O}{\|}}{C}-\overset{\overset{\displaystyle O}{\|}}{C}-CH_2-OH$$

Systematic names have the advantage that they can be decoded to reveal the structure, but for even moderately complex molecules they can be very long and clumsy. Non-systematic (trivial) names appear in old and modern books and other publications and are in everyday use among practising chemists and biochemists. It is wise to know both names for important substances such as ethanoic acid (acetic acid). Both types of name appear in later chapters and the Index will provide a 'translation' when necessary.

## SHORTHAND DRAWINGS

It is often convenient to adopt a more shorthand way of drawing molecules which omits more bonds then we have done so far. Thus pent-2-en-1-ol could be written as $CH_3CH_2CH=CHCH_2OH$. The only bond which is put in explicitly is the double bond. Likewise, 3-methylbutanoic acid could be written as $(CH_3)_2CHCH_2COOH$ and the aldehyde 3-amino-4-chlorobutanal as $ClCH_2CH(NH_2)CH_2CHO$. The bonding arrangements in the carboxyl group and in the aldehyde group are, of course, as drawn in detail above.

The in-line drawing of carboxyl groups often causes confusion. By convention the oxygen of the carbonyl group should always be drawn immediately to the right of its carbon atom. Thus $CH_3COOH$ and $HOCOCH_3$ are both acceptable ways of drawing ethanoic acid (acetic acid), and $CH_3CH_2OCOCH_3$ and $CH_3COOCH_2CH_3$ both represent the ethyl ester of this acid, its full structure being

$$CH_3-C\!\!\begin{array}{c}\nearrow O \\ \searrow O-CH_2-CH_3\end{array}$$

An even more simplified method of drawing molecules is to draw only the skeletal bonds, to omit all H and C atoms and to assume that there is a carbon atom at each bond end and at each bond junction and that every carbon atom has enough hydrogens attached to satisfy its valency of four:

represents

$$\begin{array}{c}CH_2-CH_2\diagdown\quad\diagup CH_3 \\ \qquad\qquad C \\ CH_2-C\diagup\quad\diagdown CH_3 \\ \quad\;\diagup\;\diagdown \\ \quad H\;\; OH\end{array}$$

2,2-dimethylcyclopentan-1-ol

represents

$$CH_3CH_2CH_2CH_2-C\!\!\begin{array}{c}\diagup\overset{H}{C}\diagdown \\ \diagup\quad C\diagup\diagdown O \\ H_2C\diagdown\quad\diagup CH-CH_3 \\ \qquad C \\ \qquad H_2\end{array}$$

3-butyl-6-methylcyclohex-2-en-1-one

Benzene and compounds related to it can be drawn in the same way.

Benzene

1,4-Dimethylbenzene

1-Phenylpropan-2-one

## STRUCTURAL ISOMERS

It is clear that, from a small assembly of atoms, several different molecules could be constructed. Two quite different molecules with different functional groups have the molecular formula $C_2H_6O$, viz. $CH_3CH_2OH$ and $CH_3OCH_3$. The same set of atoms are bonded together differently in the two cases. The two molecules have different *structures* but the same molecular formula. They are *structural isomers*. There are three structural isomers with the molecular formula $C_2H_4O$:

$$CH_3-C\overset{\displaystyle O}{\underset{\displaystyle H}{\big\langle}} \qquad CH_2=C\overset{\displaystyle OH}{\underset{\displaystyle H}{\big\langle}} \qquad \text{and} \qquad \underset{O}{\overset{CH_2-CH_2}{\triangle}}$$

2-Chloropropan-1-ol and 1-chloropropan-2-ol are also structural isomers. They are different molecules with different physical and chemical properties, even though they have the same functional groups.

*Problem 3–4. There are several further structural isomers with the molecular formula $C_3H_7ClO$. Not all of them are alcohols and one is a salt. Not all of them have been made. Can you find seven more?*

In all cases the systematic name of a substance defines its structure and it is important to be able to interpret systematic names of molecules. However, there are more important topics in organic chemistry than the niceties of nomenclature. One of these is the shapes of molecules, and these are discussed in the next chapter.

## ANSWERS TO PROBLEMS IN CHAPTER 3

*Problem 3–1.*

$CH_4 \quad CH_3CH_3 \quad CH_2=CH_2 \quad HC\equiv CH \quad {}^{\oplus}CH_3 \quad {}^{\ominus}\bar{C}H_3$

$CH_3Li \quad CH_3BeCH_3 \quad CH_3BeF$

$CH_3BH_2 \quad (CH_3)_2BH \quad CH_3CH_2BH \quad \underset{\overset{|}{B}\atop{\overset{|}{H}}}{\overset{CH_2-CH_2}{\triangle}} \quad CH_3\overset{\ominus}{B}H_3 \quad \text{etc.}$

$CH_3\bar{N}H_2 \quad (CH_3)_2\bar{N}H \quad CH_2=CH-\bar{N}H_2 \quad (CH_3)_2\overset{\ominus}{N}\bar{\mathrm{I}} \quad CH_3\overset{\oplus}{N}H_3 \quad CH_2=\bar{N}H \quad \underset{\overset{|}{N}\atop{\overset{|}{H}}}{\overset{CH_2-CH_2}{\triangle}}$

$\overset{\oplus}{C}H_2=\bar{N}H_2 \quad {}^{\ominus}\bar{C}\equiv\bar{N} \quad CH_3-C\equiv\bar{N} \quad CH_3-\overset{\oplus}{N}\equiv\bar{C}{}^{\ominus}$

${}^{\ominus}\bar{C}=\bar{O}{}^{\oplus} \quad \bar{O}=C=\bar{O} \quad CH_3-\bar{O}-H \quad CH_3-\bar{O}-CH_3 \quad CH_3CH_2\bar{O}H \quad CH_2=CH-\bar{O}-H$

$\underset{\bar{O}}{\overset{CH_2-CH_2}{\triangle}} \quad CH_2=\bar{O} \quad CH_3-\underset{\overset{|}{H}}{C}=\bar{O} \quad CH_3-\overset{\oplus}{O}-H \text{ etc.} \quad CH_3-\bar{O}\mathrm{I}^{\ominus} \text{ etc.} \quad H_2\overset{\oplus}{C}-\bar{O}-H \text{ etc.}$

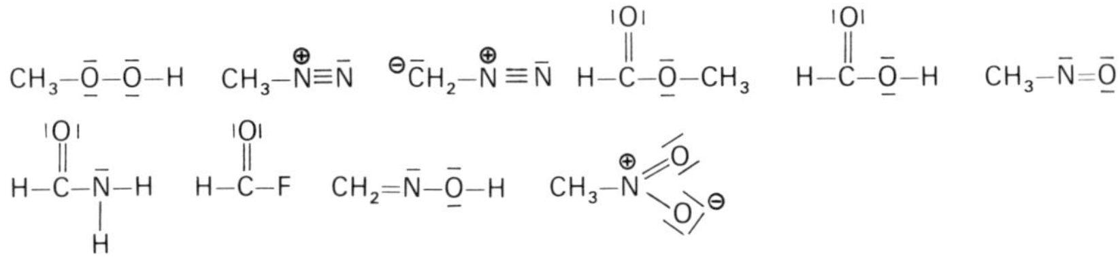

$$CH_3-\overset{..}{\underset{..}{O}}-\overset{..}{\underset{..}{O}}-H \quad CH_3-\overset{\oplus}{N}\equiv\overset{..}{N} \quad \overset{\ominus}{\overset{..}{C}}H_2-\overset{\oplus}{N}\equiv\overset{..}{N} \quad H-\overset{\overset{\displaystyle |\overset{..}{O}|}{\|}}{C}-\overset{..}{\underset{..}{O}}-CH_3 \quad H-\overset{\overset{\displaystyle |\overset{..}{O}|}{\|}}{C}-\overset{..}{\underset{..}{O}}-H \quad CH_3-\overset{..}{N}=\overset{..}{\underset{..}{O}}$$

$$H-\overset{\overset{\displaystyle |\overset{..}{O}|}{\|}}{\underset{\underset{\displaystyle H}{|}}{C}}-\overset{..}{N}-H \quad H-\overset{\overset{\displaystyle |\overset{..}{O}|}{\|}}{C}-F \quad CH_2=\overset{..}{N}-\overset{..}{\underset{..}{O}}-H \quad CH_3-\overset{\oplus}{N}\underset{\overset{\displaystyle \diagdown O}{}\ominus}{\overset{\displaystyle \diagup O}{}}$$

$$CH_3F \quad CH_2F_2 \quad CHF_3 \quad CH_2=CHF \text{ etc.}$$

Not all of these substances are known and some are very reactive. Some will appear again later in this book.

*Problem 3–2.*
alcohol   alkyl halide   alkene   ether   amine
aldehyde   carboxylic acid   ester   amide   nitroalkane
alcohol   amine   alkyne

*Problem 3–3.* From left to right: 2-methylbut-1-ene; 3-amino-2-methylpropanoic acid; and butane-2,3-dione-1,4-diol or 1,4-dihydroxybutane-2,3-dione.

*Problem 3–4.*

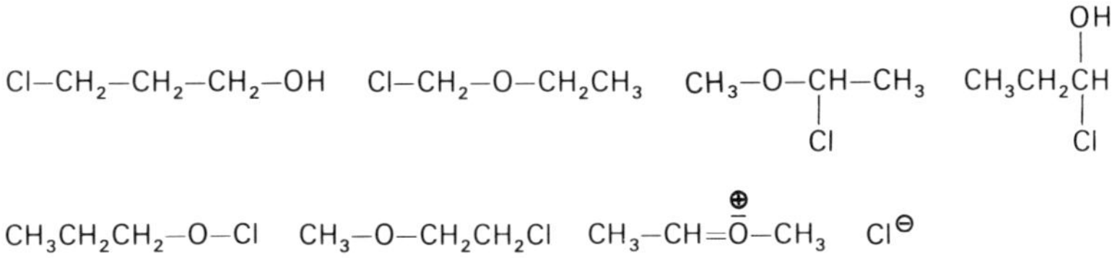

$$Cl-CH_2-CH_2-CH_2-OH \quad Cl-CH_2-O-CH_2CH_3 \quad CH_3-O-\underset{\underset{\displaystyle Cl}{|}}{CH}-CH_3 \quad CH_3CH_2\underset{\underset{\displaystyle Cl}{|}}{\overset{\overset{\displaystyle OH}{|}}{CH}}$$

$$CH_3CH_2CH_2-O-Cl \quad CH_3-O-CH_2CH_2Cl \quad CH_3-CH=\overset{\oplus}{O}-CH_3 \quad Cl^{\ominus}$$

# Chapter 4

## Stereochemistry

The prefix stereo- comes from a Greek word meaning solid or three dimensional. What other modern words contain this prefix? This chapter is concerned, therefore, with the shapes of molecules.

Previous chapters discussed the structures of molecules, that is, the sequence in which atoms are joined to each other, but did not discuss shapes. We might ask a series of questions. How should we describe the shape of a molecule? How easily and in what ways can the shape be changed by changes in temperature or environment (solvent, crystalline state)? How can we find this information? Why has the molecule this shape? These are penetrating questions and in this chapter we can only begin to provide answers.

### DESCRIBING MOLECULAR SHAPE

To describe the shape of a molecule such as water whose structure is H—O—H we need to know two measurements only—the H—O distance and the H—O—H bond angle. Note that it is assumed (correctly) that the two H—O distances (the bond lengths) are identical and that the positions of the nuclei only are being considered. The electrons have been ignored.

For a more complicated molecule such as hydrogen peroxide with the structure H—O—O—H we need to know the bond lengths (H—O and O—O) and the H—O—O bond angle, but that is not sufficient. Even if we know these, we do not know whether all four atoms lie in a plane as in A or B below or whether the molecule has a three-dimensional spiral shape with one hydrogen atom not lying in the plane defined by the other three atoms:

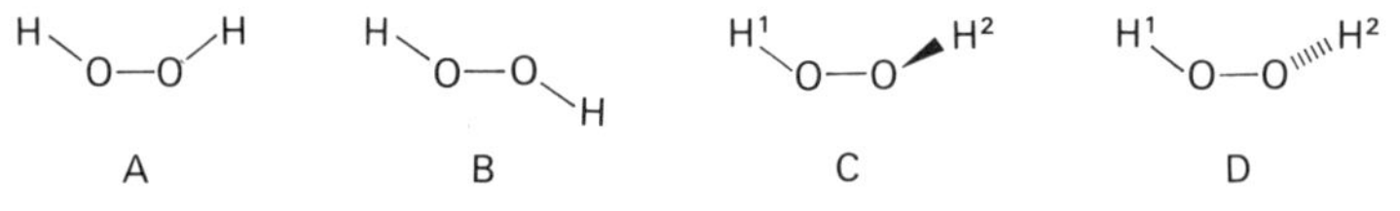

$$
\begin{array}{cccc}
\text{A} & \text{B} & \text{C} & \text{D}
\end{array}
$$

The wedge in C is used to indicate that $H^2$ is lying in front of the $H^1OO$ plane, and the dashed line in D to show it lying behind. The amount of twist can be described by the angle between the $H^1OO$ plane and the $H^2OO$ plane. This angle is called a *dihedral angle* or torsion angle. The molecule could be envisaged as lying in a partly open book with the oxygen atoms on the spine and one hydrogen atom on each page. The dihedral angle is then the angle between the pages.

Alternatively, if you line up the molecule so that you are looking along the O—O bond then the molecule might look like

$$H^1$$
$$|$$
$$O\ \,^{-H^2}$$

The other oxygen atom is directly behind the first. The dihedral angle is now seen as the $H^1$—O—$H^2$ angle in this projection.

For three-dimensional molecules containing four or more atoms in a row there will be dihedral angles, whose value we would need to know, for every set of four atoms joined as

$$W\diagdown_{X}\diagup^{Y}\diagdown_{Z}.$$

If we know all the bond lengths, bond angles and dihedral angles we know the shape of the whole molecule.

If in the hydrogen peroxide molecule the dihedral angle is gradually increased from 0° to 180° with no changes in bond lengths or bond angles, the effect is to rotate one end of the molecule relative to the other round an axis which runs through the two oxygen atoms. This is usually called rotation about the O—O bond. The shape of the molecule will change from structure A above through C to B. Further rotation will give D and eventually (after 360°) A again. The different shapes A, C, B, and all the intermediate ones are called different *conformations* of the molecule. The conformations of a molecule are therefore the shapes it might have which differ by changes in dihedral angles only, that is, differ by rotations about bonds. Hydrogen peroxide has an infinite number of possible conformations.

## ELASTICITY OF BOND LENGTHS

If we chose to define the shape of the molecule in terms of bond lengths, bond angles and dihedral angles, we should discuss deformations of the molecule as changes in these three measurements. Small changes are occurring all the time as the molecule vibrates but we are concerned here with average bond lengths and angles. It is found that changes in bond lengths from their usual values require very considerable increases in energy for a small percentage change in length.

One result of the 'stiffness' of covalent bonds is that a given bond, say a $C-C$ single bond has about the same length, approximately 153 pm = $153 \times 10^{-12}$ m = 1.53 Å, no matter what its surroundings are. $C-C$ bond distances of up to 156 pm have been found, i.e. 2% above average.

Some average bond lengths (in pm) are as follows:

|       |     |        |     |        |     |       |     |
|-------|-----|--------|-----|--------|-----|-------|-----|
| $C-C$ | 153 | $C-N$  | 147 | $C=C$  | 133 | $O-H$ | 96  |
| $C-O$ | 142 | $C-Cl$ | 177 | $C=O$  | 121 | $C-H$ | 109 |

## ELASTICITY OF BOND ANGLES

Changes in bond angles also require large increases in energy for small changes of angle. For carbon atoms forming bonds to four other atoms, as in $CH_4$ or $CH_3-CH_2-CH_3$, the bond angles round the carbon atom are usually in the range 108–112°. Molecules with angles greater or smaller than this will be strained and will exhibit abnormal reactivity. Thus cyclopropane, which has $C-C-C$ bond angles of 60°, undergoes reactions which cyclopentane does not. Reactions which result in ring opening of cyclopropane will allow relief of the angle strain, i.e. will allow a reduction in the energy of the molecule.

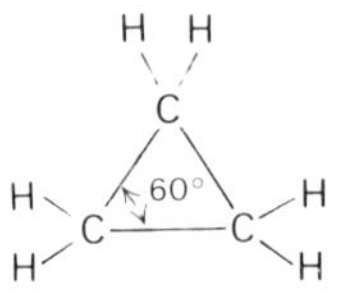

Cyclopropane

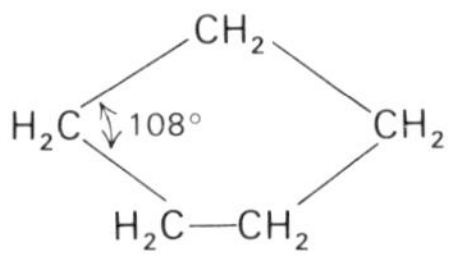

Cyclopentane

## CHANGES IN DIHEDRAL ANGLES

The changes in dihedral angles involved in rotations about single bonds are, however, fairly easy. The energy differences between conformers are usually less than the differences in kinetic energy between different molecules in a sample at room temperature. At room temperature, therefore, there is essentially free

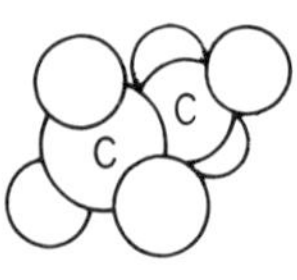

Ethane, $C_2H_6$, Rotation about the $C-C$ bond is easy

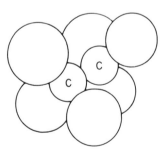

Hexachloroethane, $C_2Cl_6$. Rotation about the $C-C$ bond is difficult

rotation about all single bonds in acyclic molecules. The situation would change if the temperature were reduced or if the molecule possessed a ring preventing rotation or if the bond were not a single bond (p. 36), or if the substituents were bulky as in $Cl_3C-CCl_3$.

What is important here is that it is usually not difficult for a molecule to adopt any of its conformations if this is required for, say, adsorption or reaction, whereas it is difficult for it to change its bond lengths or bond angles by much.

## SHAPES FOUND FOR REAL MOLECULES

The shapes and the elasticity of many molecules have been studied by a variety of spectroscopic methods. The results observed for bond angles can be summarized as follows. The atoms or groups round a central atom are as far apart as possible, that is, the angles are all as large as possible. This is true even if the bond attaching the group to the central atom is a multiple (double or triple) bond. If there are lone pairs of electrons, each such pair must be considered as a space-requiring group for this purpose, just like the bonding pairs and their attached atoms. Thus ammonia, $\overline{N}H_3$, would be expected to have much the same H–N–H bond angles as the ammonium ion, $NH_4^{\oplus}$, since both have four 'groups' (in the present sense of the word) round the nitrogen atom.

If you use a ball of plasticine and some sticks to explore the arrangements $AX_2$, $AX_3$, $AX_4$, etc., in order to find what shape the assembly has when the groups X are furthest apart, the results turn out to be simple symmetrical figures, as shown in the table below.

| Formula | Diagram | Description | X–A–X angle |
|---------|---------|-------------|-------------|
| $AX_2$ | X–A–X | Linear | 180° |
| $AX_3$ | X–A$\langle^X_X$ | Trigonal (planar) | 120° |
| $AX_4$ | $X_{\prime\prime\prime}$A$^X_X$ X | Tetrahedral | 109° 28′ |
| $AX_6$ | $X_{\prime\prime\prime}$A$^X_X$ with X top and bottom | Octahedral | 90° |

In these four cases, there is a unique answer, a highly symmetrical figure in which the environment of each group X is the same within each figure. If the groups X were joined by lines we would form in the $AX_3$ case an equilateral triangle; in the $AX_4$ case a tetrahedron, that is a pyramid with four faces each an equilateral triangle; and in the $AX_6$ case an octahedron or double pyramid with eight equilateral triangular faces.

The $AX_5$ case is more tricky and less symmetrical. It is relevant in organic chemistry only for the transition states of some reactions. The $AX_6$ case is

FERNALD LIBRARY
COLBY-SAWYER COLLEGE
NEW LONDON, N.H. 03257

34

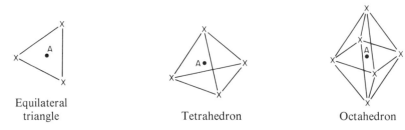

| Equilateral triangle | Tetrahedron | Octahedron |

important for inorganic compounds. The $AX_4$, $AX_3$, and $AX_2$ cases are all important in organic chemistry. Note that in the tetrahedral $AX_4$ arrangement the angle is about $109\frac{1}{2}°$ and a plane drawn through A and two of the groups X is such that the remaining X groups are mirror images of one another in that plane.

## TETRAHEDRAL MOLECULES

Actual molecules such as $CH_4$, $CCl_4$, $NH_4^\oplus$, $BH_4^\ominus$, and $Si(CH_3)_4$ have four identical groups round a central atom. There are no lone pairs. They are all found to be tetrahedral molecules.

In molecules such as $HCCl_3$ and $CH_2(CH_3)_2$ the four groups are not all the same. The basic tetrahedral arrangement is distorted a little to allow the larger groups to move further apart. Molecules such as $(HO)_3P{=}O$, orthophosphoric acid, which contains a double bond, are also almost perfectly tetrahedral. Molecules such as $\bar{N}H_3$, $\bar{C}H_3^\ominus$, $\underline{O}H_2$, and $\bar{O}H_3^\oplus$ are considered to have four groups round the central atom, each lone pair being counted as a group. Therefore, the four electron pairs are arranged more or less tetrahedrally. There are small deformations since the four corners of the tetrahedron are not all occupied by identical groups.

We shall describe all of these molecules as tetrahedral, even those which do not strictly have tetrahedral symmetry.

All compounds in which a carbon atom has four groups round it are three-dimensional. Although we draw them on flat paper, we must not forget that fact. Drawings using wedges (in front) or dashed lines (behind the paper) will be used when necessary.

## USING MODELS

Molecular models are essential in order to be able to appreciate stereochemistry. 'Model' means a simplified image or representation of the real thing. Molecular models made of balls and sticks, or polypods and tubing, have fixed bond lengths and angles and usually allow rotation about bonds as in the real molecule. They do not give any idea of how much space the electrons of the atoms occupy. This can be achieved by using models made of spheres of appropriate sizes with rubber press-in connecting pieces. These usually are more difficult to move. Study the different types of models available to you to see in how much they behave like real molecules.

Make a model of ethane and study its conformations. Make a model of hexane. All kinds of curled conformations are possible. In the solid state the molecules adopt a zig-zag arrangement.

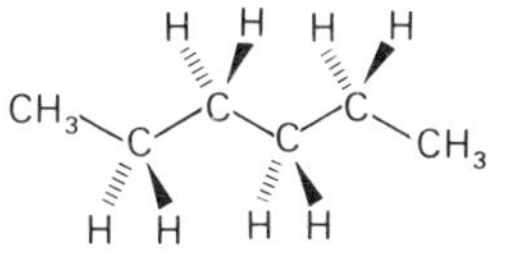

One of the conformations of hexane

When we speak of unbranched chains as 'linear', we do not imply that the carbon atoms are actually in a straight line. Make a model of cyclohexane. You will find that several conformations of your model are possible. One of them should look like the picture below. This conformation is the one adopted by most molecules with a cyclohexane ring. Note that the carbon atoms are not all in one plane.

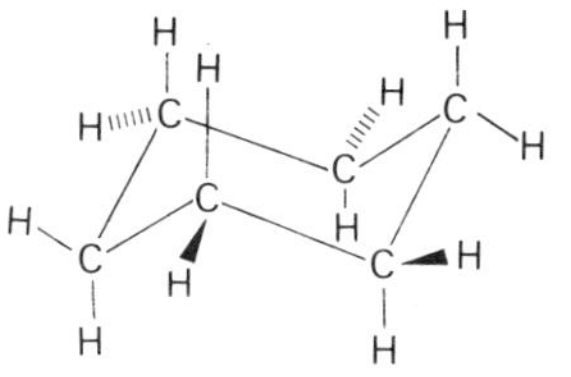

One of the conformations of cyclohexane

## TRIGONAL MOLECULES

Molecules with three groups X, such as $BF_3$ (boron trifluoride) and $CH_2O$ (formaldehyde, methanal), are planar and have angles close to 120°. We shall describe them all as trigonal, even those which do not possess strict trigonal symmetry. Although there is a C=O double bond in formaldehyde, the oxygen is considered as one group. The H—C—O angle is very close to 120°. Carbonium ions of the type $^{\oplus}CX_3$ are also planar and trigonal.

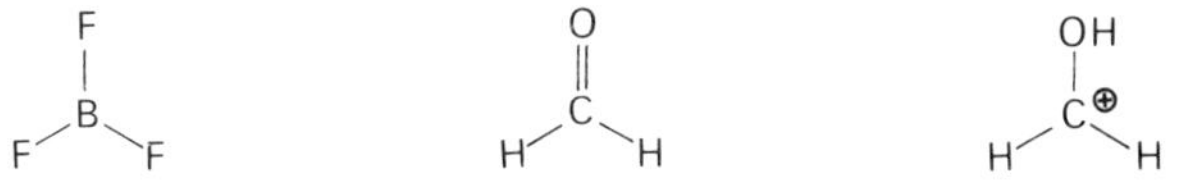

*Problem 4–1. What shapes would you expect the following molecules or ions to be: $PH_3$, $BO_3^{3-}$, $CO_3^{2-}$, $PO_3^{3-}$, $HO_2^{\ominus}$, $HC_2^{\ominus}$, $NO_2^{\ominus}$, $BeCl_2$, $SO_2$, $NCO^{\ominus}$, $SF_6$, $BF_4^{\ominus}$, $CH_2F_2$?*

## MOLECULES WITH DOUBLE BONDS

What will happen if a molecule contains a double bond joining two adjacent trigonal atoms? In ethene both carbons are trigonal. The atoms C, C, $H^1$ and $H^2$ must all be in a plane and C, C, $H^3$ and $H^4$ must all be in one plane.

$$H^1{\diagdown} \atop {H^2}{\diagup}C{=}C{\diagup}^{H^3} \atop {\diagdown}_{H^4}$$

It is found, in fact, that *all* the atoms are in the same plane and that rotation of one end of the molecule relative to the other requires a considerable increase in the energy of the system. Rotation about double bonds, unlike rotation about single bonds (see p. 32), does not occur at room temperature. The energy of the twisted conformers is so much higher than that of the planar normal state that no molecules in the population have enough energy at room temperature to get into the conformation with a dihedral angle of 90°. The same is true for other doubly bonded molecules.

$$\begin{array}{ccc} {H{\diagdown} \atop H{\diagup}}C{=}{\overset{\oplus}{N}}{{\diagup}H \atop {\diagdown}H} & {H{\diagdown} \atop H{\diagup}}C{=}N{\diagdown}_{OH} & {H{\diagdown} \atop H{\diagup}}C{=}{\overset{\oplus}{O}}{\diagdown}_{H} \end{array}$$

What effects will this have on molecular flexibility? Make models of butane and but-2-ene. Do the models you have show the differing ease of rotation about the central bond in the two molecules? How many different but-2-enes did you find? How are they related? Will they be interconvertible at room temperature? They are discussed again below.

How many but-2-ynes are possible? What shape is this molecule? Does it fit the generalization above? Carbon atom number two in the chain has two attached groups, a $-CH_3$ and a $\equiv CCH_3$. It should therefore be linear. So must be C-3, so the whole carbon skeleton is rigorously linear (unlike that of the 'linear' alkanes above).

If a molecule contains a double bond which has two non-identical groups at both its ends, there will be two different forms of the molecule which do not isomerize, the one into the other, at room temperature. The two but-2-enes have the same structure but different stereochemistries. The prefixes *cis* and *trans* are used to distinguish them. The *cis* isomer is defined as the one in which the larger substituent at one end is on the *same* side of the molecule as the larger of the substituents at the other end.

$$\begin{array}{ccc} {CH_3{\diagdown} \atop H{\diagup}}C{=}C{{\diagup}CH_3 \atop {\diagdown}H} & {CH_3{\diagdown} \atop H{\diagup}}C{=}C{{\diagup}H \atop {\diagdown}CH_3} & {H{\diagdown} \atop Cl{\diagup}}C{=}C{{\diagup}CH_3 \atop {\diagdown}H} \end{array}$$

*cis*-but-2-ene          *trans*-but-2-ene          *trans*-1-chloropropene

The same phenomenon occurs with 1-chloropropene and with 3-chloropent-2-enoic acid, although in the latter case the naming of the isomers would need more care. Nomenclature rules for such difficult cases have been devised which define 'larger' more precisely.

$$CH_3CH_2 \diagdown C=C \diagup COOH \atop Cl \diagup \diagdown H$$   $$CH_3CH_2 \diagdown C=C \diagup H \atop Cl \diagup \diagdown COOH$$

The two isomeric 3-chloropent-2-enoic acids

These stereoisomeric pairs of molecules are called *geometric isomers* since they differ in shape. They have different physical and chemical properties. The melting and boiling points of *cis*- and *trans*-but-2-enoic acids are given below:

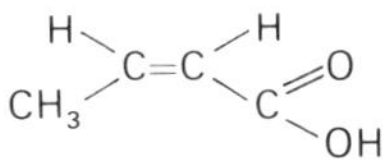

$$H \diagdown C=C \diagup H \atop CH_3 \diagup \diagdown C \diagdown_{OH}^{O}$$   $$CH_3 \diagdown C=C \diagup H \atop H \diagup \diagdown C \diagdown_{OH}^{O}$$

m.p. 16 °C, b.p. 169 °C 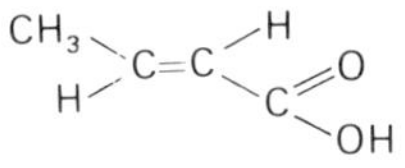      m.p. 72 °C, b.p. 189 °C

Molecules such as 1,1,2-trichloroethene do not, of course, have two forms:

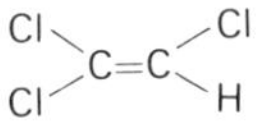

$$Cl \diagdown C=C \diagup Cl \atop Cl \diagup \diagdown H$$

Trichloroethene

## MOLECULES WITH RINGS

The term geometric isomers and the labels *cis* and *trans* are also used for another kind of stereoisomer. Make models of 1,2-dimethylcyclopentane. You can make one in which the methyl groups are on the same face of the ring and one in which they are on opposite faces.

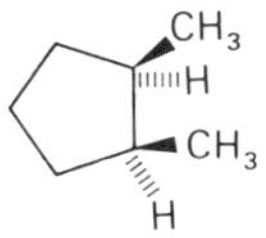

*cis*-1,2-Dimethylcyclopentane 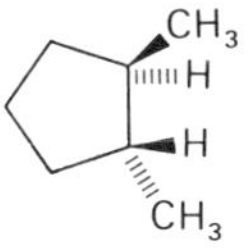      *trans*-1,2-Dimethylcyclopentane

You may have found that you can make two different *trans*-1,2-dimethylcyclopentanes. We shall return to that intriguing problem later.

Geometric isomerism can occur in cyclic compounds no matter how big the ring is.

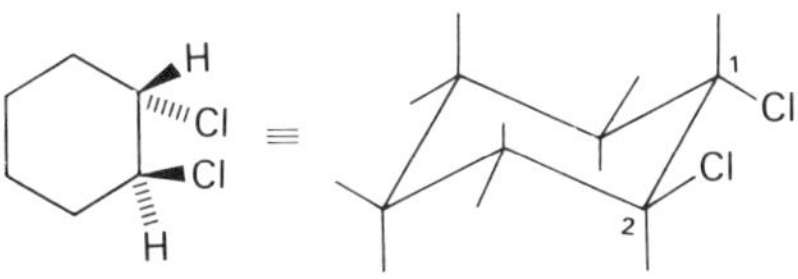

*trans*-1,2-Dichlorocyclohexane 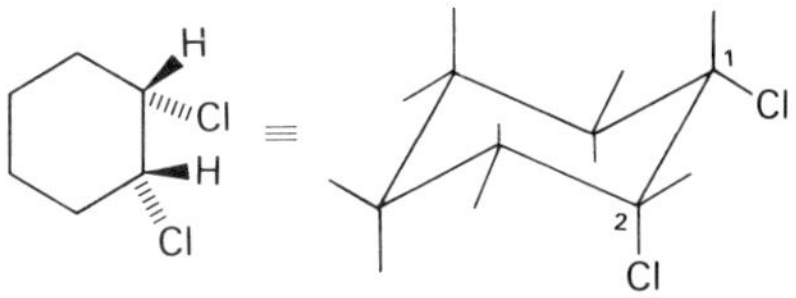      *cis*-1,2-Dichlorocyclohexane

In the first drawing of *cis*-1,2-dichlorocyclohexane both chlorine atoms are drawn with dashed bonds, indicating that they are both on the same side of the

38

ring (both behind or below the ring). In the drawing of the conformation which the molecule adopts this is still true. The chlorine at C-1 is the lower of the two groups (H and Cl) attached to the ring and at C-2 the chlorine is the lower of the two groups; that is, the chlorines are *cis*

In the drawing of the conformation adopted by *trans*-1,2-dichlorocyclohexane the chlorine is the lower substituent at C-1 and the chlorine is the upper substituent at C-2. The chlorines are *trans*.

*Problem 4–2. Study diagrams or models of pent-2-ene, 2-methylbut-2-ene and hexa-2,4-diene. How many geometric isomers are there in each case?*

*Problem 4–3. Study diagrams or models of 1,3-dimethylcyclopentane, 1,1,2-trimethylcyclopentane and 1,2,3-trimethylcyclopentane. How many geometric isomers are there in each case?*

## CHIRALITY—DO YOU KNOW YOUR LEFT HAND FROM YOUR RIGHT?

Make a model of the tetrahedral molecule CHFClBr using different coloured balls or sticks to represent the four different groups attached to the carbon. Without destroying this model, make a second model which is the mirror image of the first. Are the models identical? Can you lift one and set it on top of the other so that the balls of the same colours are touching? Both models are models of bromochlorofluoromethane. Both have the same bond lengths and angles, but they are not identical. They are not superposable.

The bromochlorofluoromethane molecule can therefore have two forms. Try making other models of the molecule by rearranging the attached groups. Are there any further possibilities? All of the attempts turn out to be the same as the first model or as the second one which is the mirror image of the first.

We usually assume that the face we see when we look in a mirror is an exact replica of our own. Often it is; but try shutting your right eye. Is it the right eye of the face in the mirror which shuts?

You may have seen adverts like

# ƨIXAT ƧᙠOᙠ

written on the fronts of vans. The mirror image of this message is not the same as the original.

Some objects such as balls, road rollers and cigarettes have mirror images exactly the same as themselves. Can you think of others? But some objects, such as motor cars and spiral stairs, have mirror images which have the same dimensions as the original but are not identical.

All objects can be divided into two classes, those whose mirror image is not identical with the original and those whose mirror image is identical with the original. Hands, shoes, screws, coil springs, propellers, and golf clubs are

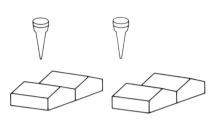

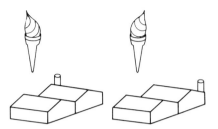

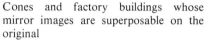

Cones and factory buildings whose mirror images are superposable on the original

Cones and factory buildings whose mirror images are not superposable on the original

examples of the first class, which are called *chiral* objects (pronounced kyral). Balls, pencils, tables, Bunsen burners, shoe boxes, and teddy bears belong to the second class, which are *non-chiral* or *achiral* objects. Strictly, the classification of teddy bears depends on their stance. The classification of Bunsen burners depends on how the tubing is bent and how the collar is set. Since each of these objects can be moved into a stance in which it has a mirror image identical with itself we can label them as non-chiral.

Teddy pointing,
chiral

Teddy sunbathing,
non-chiral

Molecules can be classified likewise. Bromochlorofluoromethane is a chiral molecule. Methane is non-chiral. Hydrogen peroxide is non-chiral since it is able to adopt at least one conformation which is non-chiral.

$$H\diagdown_{O-O}\diagup^{H}$$

A non-chiral conformation
Dihedral angle (see p. 31) = 0°

$$H\diagdown_{O-O}^{\prime\prime\prime\prime H}$$

A chiral conformation
Dihedral angle = say 60°

The word chiral comes from the Greek word for a hand, which appears also in the words chiropodist and chirurgical. A hand is a conveniently demonstrable chiral object. Chiral molecules have a handedness or chirality. They exist in two forms which are *enantiomeric* that is, which are non-identical, non-superposable, mirror images. The one is the enantiomer of the other. The enantiomeric forms of a molecule are stereoisomers. They differ only in the three-dimensional arrangement of the structure. They are sometimes called optical isomers for a historical reason, which we shall not discuss here. A mixture of equal numbers of the two enantiomeric forms of a substance is called a *racemic mixture* or a *racemate*.

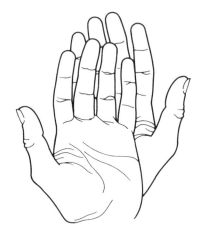

## CHIRALITY AND SYMMETRY

What kinds of objects are non-chiral? Can you detect any property common to the non-chiral objects in the list above? Non-chiral objects have some kind of symmetry which chiral objects do not have. An interesting example is that of a pair of shoes:

A pair of shoes assembled in a non-chiral manner with a plane of symmetry

A pair of shoes assembled in a non-chiral manner with a centre of symmetry

A pair of shoes assembled in a chiral manner

A pair of shoes laid side by side form an object whose mirror image (another pair) is identical with the original. The pair arranged in this way is a non-chiral object. It possesses a *plane of symmetry*. A plane of symmetry is a plane which can be imagined to run through an object dividing it in halves which are mirror images of one another. Balls, road rollers, pencils and sunbathing teddy bears all have at least one plane of symmetry.

A pair of shoes laid one upside down parallel to the other, the way they are put in a shoe box, also constitutes a non-chiral object. It possesses a *centre of symmetry*. A centre of symmetry is a point in the middle of an object such that for every part of the object there is an identical part equidistant from that point and on the other side of it. Thus four pairs of coloured balls stacked so that they are at the corners of a cube as shown below form an assembly which has a centre of symmetry at the centre of the cube. This assembly is non-chiral but it has no plane of symmetry.

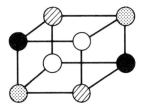

A pair of shoes standing heel to toe in line has no symmetry. It is said to be asymmetric and is a chiral object. Objects such as hands, screws, and propellers have no planes or centres of symmetry. They are, therefore, chiral. Note that the propeller has an axis of symmetry. If it has three blades it could be rotated round its spindle by 120° and would then look exactly as it did originally. But the propeller cannot be superposed on its mirror image. In spite of the axis of symmetry, it is chiral. Axes of symmetry do not affect chirality.

To sum up, an object is chiral if it is not superposable on its mirror image and this is the best test for chirality. If the object has a plane or a centre of symmetry we know that its mirror image will be superposable on the original and the object will be non-chiral.

In the same way, any molecule possessing (in any of its accessible conformations) a plane of symmetry or a centre of symmetry will be non-chiral. It follows immediately that planar molecules such as methanal and chloroethene are non-chiral since the plane of the nuclei is a plane of symmetry.

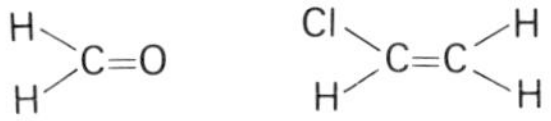

$$\begin{array}{c} H \\ \diagdown \\ H \diagup \end{array} C{=}O \qquad \begin{array}{c} Cl \\ \diagdown \\ H \diagup \end{array} C{=}C \begin{array}{c} \diagup H \\ \diagdown H \end{array}$$

Planar non-chiral molecules

*Problem 4–4. Which of the following molecules are chiral? If you cannot reach a conclusion by inspecting the pictures make models and compare them.*

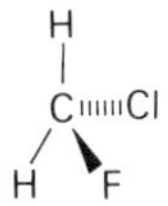

$$\underset{\substack{\text{H} \\ | \\ \text{C} \text{\tiny IIIII} Cl \\ \diagup \quad \diagdown \\ \text{H} \quad \text{F}}}{}$$

$$\underset{\substack{CH_3 \\ | \\ C \text{\tiny IIIII} CH_3 \\ \diagup \quad \diagdown \\ Cl \quad F}}{}$$

$$\underset{\substack{CH_3 \\ | \\ C \text{\tiny IIIII} CH_2F \\ \diagup \quad \diagdown \\ H \quad Cl}}{}$$

*Chlorofluoromethane* 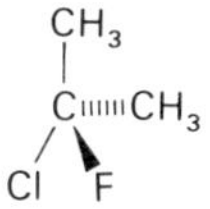   *2-Chloro-2-fluoropropane* 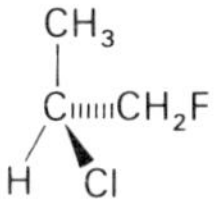   *2-Chloro-1-fluoropropane*

Tetrahedral molecules, $X_{abcd}$, in which the four groups are all different are chiral.

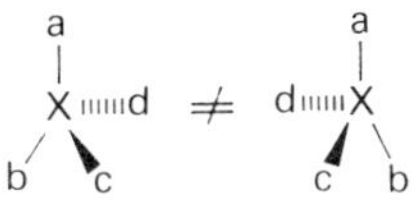

$$\underset{\substack{a \\ | \\ X \text{\tiny IIIII} d \\ \diagup \quad \diagdown \\ b \quad c}}{} \neq \underset{\substack{a \\ | \\ d \text{\tiny IIIII} X \\ \diagup \quad \diagdown \\ c \quad b}}{}$$

If two or more of the groups are identical, as in $X_{aabc}$, the molecule has a plane of symmetry and so is non-chiral. Molecules such as lactic acid, alanine, and 2-bromobutane are simple and important examples of chiral molecules, each of which possesses one tetrahedral assembly of the type $X_{abcd}$.

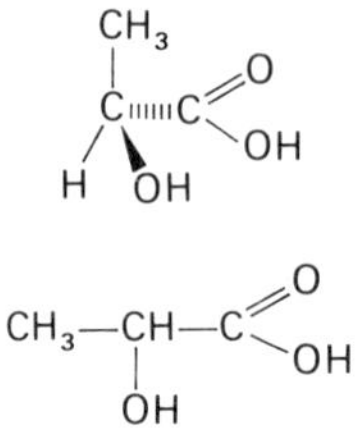

Lactic acid 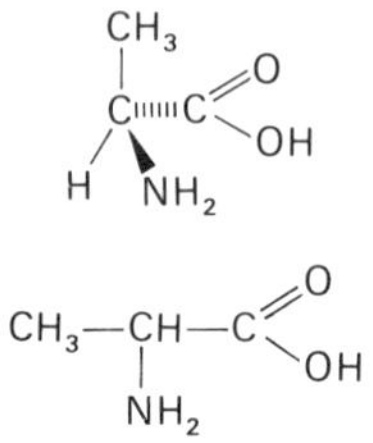          Alanine 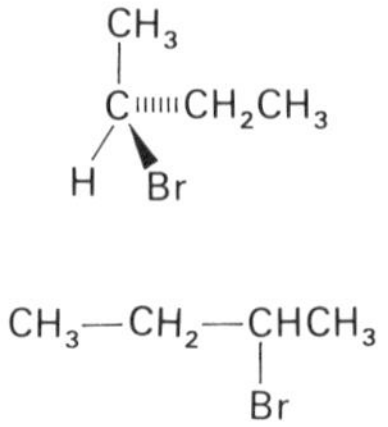          2-Bromobutane

*Problem 4–5. Which of the following molecules are chiral?*

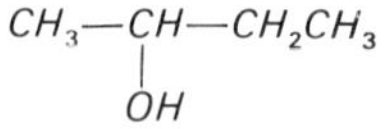

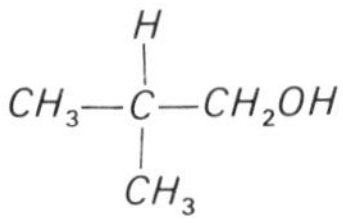

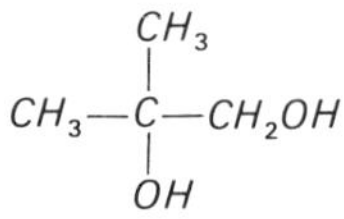

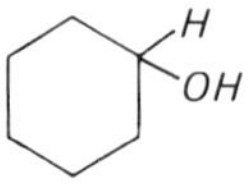

$CH_3$—$CH$—$CH_2CH_3$
|
$OH$

$CH_3$—$C$—$CH_2OH$
with $H$ above and $CH_3$ below

$CH_3$—$C$—$CH_2OH$
with $CH_3$ above and $OH$ below

Cyclohexanol 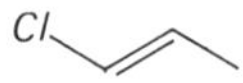          1-Chloropropene 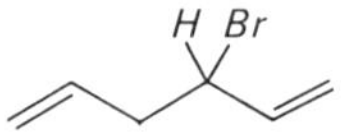          3-Bromohexa-1,5-diene

Tetrahedral molecules are not the only type which can be chiral. Suitably substituted octahedral molecules can be chiral. For example, a molecule with the generalized formula $X_{a_2b_2c_2}$ and the shape shown is not identical with its mirror image. Try making models of such systems.

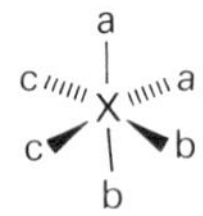

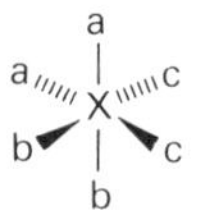

Two enantiomeric forms of $X_{a_2b_2c_2}$

The same is true of certain twisted or coiled molecules such as *trans*-cyclooctene, which can *not* adopt the planar conformation which the first drawing tends to suggest. Examination of a model shows it can adopt either of two enantiomeric, three-dimensional, twisted conformations, one of which is shown below.

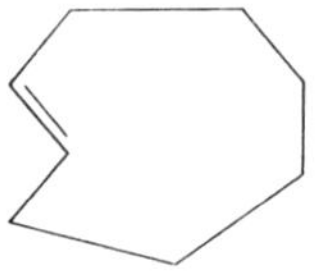

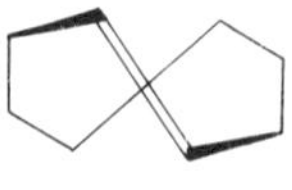

*trans*-Cyclooctene

# CHIRALITY AND BEHAVIOUR

Enantiomers are different. The difference is only slight but can be detected and may be important if the interaction of the two forms with their environment is different.

If we use a single molecule such as $X_{a_4}$ to represent the environment of another molecule $X_{pqrs}$, we can look at two cases. If the environment is non-chiral, say $X_{a_4}$, it can interact with the two enantiomers with equal success. That is, it will be possible for both of the enantiomers to meet the environment in an identical fashion.

$$p-Y\underset{q}{\overset{s}{\blacktriangleleft}}r \qquad a\overset{a}{\underset{a}{\diagdown}}X-a \qquad p-Y\underset{q}{\overset{r}{\blacktriangleleft}}s \qquad a\overset{a}{\underset{a}{\diagdown}}X-a$$

Reagent molecule     Environment     Enantiomeric reagent molecule     Environment

The complexes $X_{a_4}-Y_{pqrs}$ and $X_{a_4}-Y_{pqsr}$ are of equal energy. If, however, the environment is chiral, say $X_{abcd}$, the two complexes are not of equal energy. That is, the two enantiomers cannot meet the environment in identical ways.

$$p-Y\underset{q}{\overset{s}{\blacktriangleleft}}r \qquad b\overset{a}{\underset{c}{\blacktriangleright}}X-d \qquad p-Y\underset{q}{\overset{r}{\blacktriangleleft}}s \qquad b\overset{a}{\underset{c}{\blacktriangleright}}X-d$$

Reagent     Environment     Enantiomeric reagent     Environment

This can be easily seen by bringing up to a single hand (representing the environment) the left and then the right hand of another person. The finger tips will all 'fit' in one case but not with the enantiomeric hand.

We can conclude that enantiomers will behave identically towards non-chiral environments or reagents, but will behave differently towards chiral environments or reagents.

The enantiomeric forms of a chiral substance will therefore have different solubilities in chiral solvents, different degrees of adsorption on chiral adsorbents, different rates of movement during chromatography on chiral adsorbents, and different rates of reaction with chiral reagents, but they will have equal solubilities in non-chiral solvents, be inseparable by chromatography on non-chiral adsorbents, and react at the same rates with non-chiral reagents. They will have identical ultraviolet and infrared adsorption spectra, since the energy levels of electrons and of molecular motions are not affected by the handedness of the molecules, and they will have identical melting points and boiling points.

The intermolecular interactions in a crystal of one enantiomer are of equal energy to those in a crystal of the other. The molecules represented below as two-dimensional curls can pack in the same kind of way although, of course, the whole lattice of the second will be the mirror image of the lattice of the first.

44

ᏩᎩᎩᏩᎩᏩ
ᏩᏩᎩᏩᎩᏩ
ᏩᏩᏩᏩᏩᏩ

Two-
dimensional
diagram of a
crystal of one
enantiomer (all
righthanded
curls)

ᏋᎫᏋᎫᏋᎫ
ᏋᎫᏋᎫᏋᎫ
ᏋᎫᏋᎫᏋᎫ

Two-
dimensional
diagram of a
crystal of the
other
enantiomer (all
lefthanded
curls)

A crystal made of equal amounts of the two enantiomers can have different kinds of intermolecular attractions. This can give a different lattice unrelated to the first two. Racemates (see p. 40) may therefore have melting points different from those of the two enantiomers.

ᏋᎫᏋᎫᏋᎫ
ᏋᎫᏋᎫᏋᎫ
ᏋᎫᏋᎫᏋᎫ

Two-dimensional
diagram of a
crystal of the
racemate (equal
numbers of left
and right
handed curls)

Thus crystals of either enantiomer of lactic acid melt at 26 °C while crystals of racemic lactic acid melt at 18 °C.

When a racemic substance is mixed with a chiral reagent, the reagent may react much more slowly with one of the enantiomers than with the other. This is very important in biochemistry since the reagents or reagent–enzyme complexes are always chiral. Thus an organism may require one enantiomer of, say, the amino acid alanine (see p. 42) in its diet in order to survive, but may be poisoned if fed with the opposite enantiomer.

## SUMMARY

We have now met structural isomers, which exhibit gross differences in properties, geometric isomers, which show some differences in properties, and enantiomers, which behave differently only towards chiral reagents or environments. Conformational isomers, if they can be isolated, have different properties but usually they are so easily interconverted by rotation about single bonds that they cannot be isolated.

*Problem 4–6. How many different, that is distinguishable, alcohols are there with the formula $C_4H_{10}O$? Which are structural isomers, which are geometric isomers, and which are enantiomers?*

*Problem 4–7. How many different cycloalkanes are there with the formula $C_5H_{10}$? Which are geometric isomers, and which are enantiomers?*

*Problem 4–8. Draw all of the structurally isomeric dichloropropenes. Which of them can exist in geometrically isomeric forms? Which are chiral?*

## ANSWERS TO PROBLEMS IN CHAPTER 4

*Problem 4–1.*

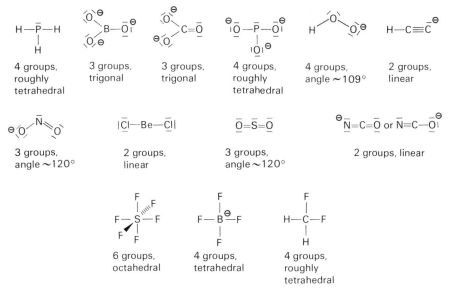

| | | | | | |
|---|---|---|---|---|---|
| 4 groups, roughly tetrahedral | 3 groups, trigonal | 3 groups, trigonal | 4 groups, roughly tetrahedral | 4 groups, angle ~109° | 2 groups, linear |

3 groups, angle ~120°

2 groups, linear

3 groups, angle ~120°

2 groups, linear

6 groups, octahedral

4 groups, tetrahedral

4 groups, roughly tetrahedral

*Problem 4–2.*

Pent-2-ene

2 ⟍⟋⟍ and ⟋⟍

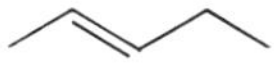

2-Methylbut-2-ene

1 ⟩=⟨

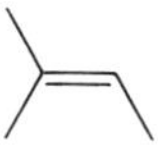

Hexa-2,4-diene

3 ⟍⟋⟍⟋ ⟍⟋⟍⟋ ⟍⟋⟍⟋

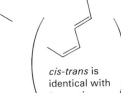

*trans-trans* 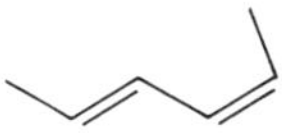   *trans-cis* 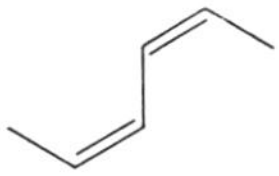   *cis-cis*   *cis-trans* is identical with *trans-cis*

46

*Problem 4–3.*

1,3-Dimethylcyclopentane

2

1,1,2-Trimethylcyclopentane

1

1,2,3-Trimethylcyclopentane

3

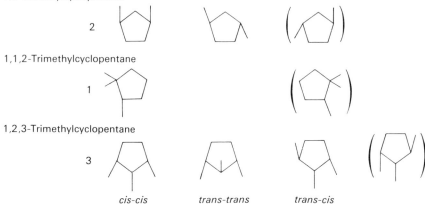

*cis-cis*          *trans-trans*          *trans-cis*

In the case of 1,3-dimethylcyclopentane there is a third isomer, which is the mirror image of the second. It is not a further geometric isomer. It has the same melting and boiling point as the second even though it is not identical with it. It is called an enantiomer of the second one. See p. 39. There is also an enantiomer of 1,1,2-trimethylcyclopentane and an enantiomer of *trans,cis*-1,2,3-trimethyl-cyclopentane.

*Problem 4–4.* Chlorofluoromethane has a plane of symmetry (the plane of the paper, as it is drawn below, in which F, C and Cl lie). It is therefore non-chiral. 2-Chloro-2-fluoropropane is also non-chiral. Both are of the type $X_{aabc}$. 2-Chloro-1-fluoropropane has no symmetry. It is chiral. Two enantiomeric forms are possible. It is of the form $X_{abcd}$.

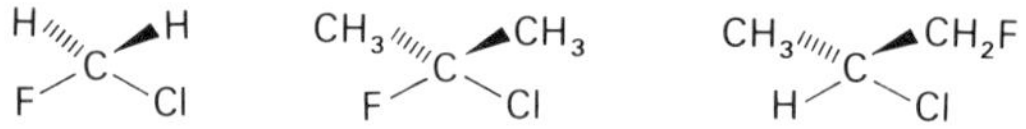

*Problem 4–5.*

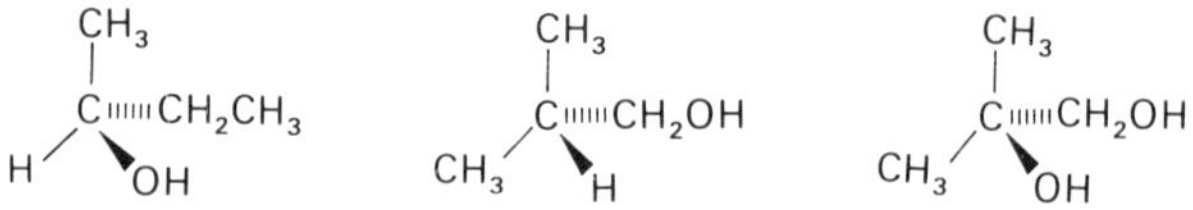

Chiral

Non-chiral

Non-chiral

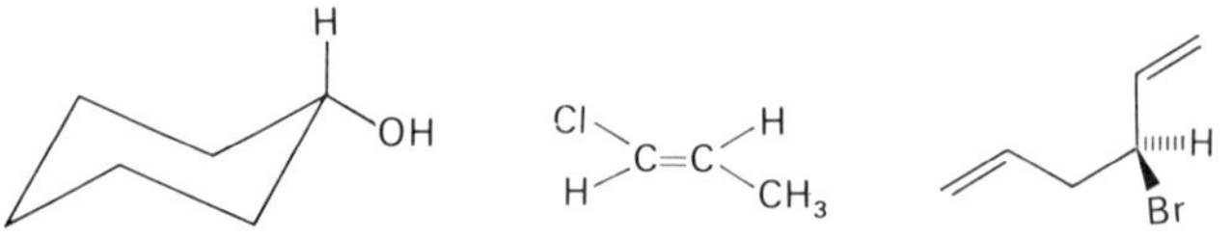

Non-chiral

Planar except for the –CH₃, non-chiral

Chiral

If you were wrong in any instance make up models and compare them. You may wish to investigate theory and practice (using models) for molecules with more than one tetrahedral centre of the type $X_{abcd}$, for example $CH_3CHBrCHBrCH_3$

and $CH_3CHBrCHClCH_3$. The dimethylcyclopentanes examined previously also fall into this class. The case of the pair of shoes is relevant here since it showed that objects made of two chiral parts *may* be non-chiral if the parts are enantiomeric.

*Problem 4–6.* There are four different structures, each with a different name:

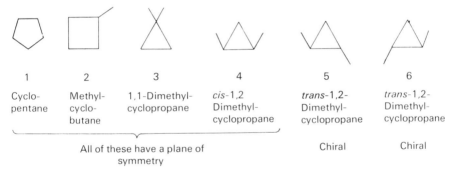

| 2-Methylpropan-2-ol | Butan-1-ol | 2-Methylpropan-1-ol | Butan-2-ol |

There are no geometrically isomeric forms possible for any of these. The last structure represents two molecules, the two enantiomeric forms of this structure. They are both called butan-2-ol. None of the other structures is chiral. Therefore there are five distinguishable alcohols, $C_4H_{10}O$.

*Problem 4–7.* There are six distinguishable cycloalkanes, $C_5H_{10}$:

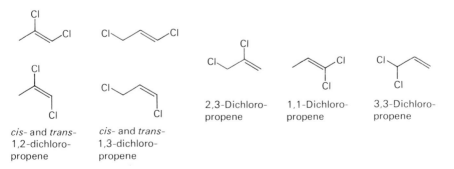

| 1 | 2 | 3 | 4 | 5 | 6 |
| Cyclo-pentane | Methyl-cyclo-butane | 1,1-Dimethyl-cyclopropane | *cis*-1,2-Dimethyl-cyclopropane | *trans*-1,2-Dimethyl-cyclopropane | *trans*-1,2-Dimethyl-cyclopropane |

All of these have a plane of symmetry

Chiral     Chiral

Compounds 4, 5 and 6 all have the same structure but different stereochemistries. Compounds 5 and 6 are enantiomers. None of the others is chiral. Compound 4 is a geometric isomer of compound 5 and of compound 6.

*Problem 4–8.* There are five structural isomers, two of which have *cis* and *trans* forms. None is chiral.

*cis*- and *trans*-1,2-dichloro-propene    *cis*- and *trans*-1,3-dichloro-propene    2,3-Dichloro-propene    1,1-Dichloro-propene    3,3-Dichloro-propene

# Chapter 5

# Electronegativity

It was assumed in Chapter 2 that the bonding electron pair in molecules such as H—H and H—Cl is shared equally by the two nuclei. This is true for the hydrogen molecule in which the ends are identical but in the molecule of HCl in the gas phase the molecule behaves as if the electrons of the bond spend more of their time nearer the chlorine nucleus than the hydrogen nucleus. The chlorine has a greater share of the bonding electrons. It has a greater attraction for the bonding electrons. The result is that the chlorine end of the molecule is slightly negative relative to the hydrogen end. The molecule is *polar*. It has two electric poles or charges. The magnitude of the charges is much less than the charge on one electron since the sharing is only slightly unequal. The symbols $\delta+$ and $\delta-$ are used to indicate these small charges:

$$\overset{\delta+\quad \delta-}{\text{H}\text{------}\text{Cl}} \qquad \text{or} \qquad \overset{\text{+}\longrightarrow}{\text{H}\text{------}\text{Cl}}$$

The term electronegativity has been coined to describe the relative ability of atoms to draw bonding electrons to themselves. The electronegativities of all of the elements can be compared and have been expressed as numbers such as those in the table below.

Table of relative electronegativities according to Pauling

| | | | | | | |
|---|---|---|---|---|---|---|
| | | H 2.1 | | | | |
| Li 1.0 | Be 1.5 | B 2.0 | C 2.5 | N 3.0 | O 3.5 | F 4.0 |
| Na 0.9 | Mg 1.2 | Al 1.5 | Si 1.8 | P 2.1 | S 2.5 | Cl 3.0 |
| | | | | | | Br 2.8 |
| | | | | | | I 2.5 |

Various different definitions and different procedures for evaluating electro-negativities all agree on general trends. Elements at the top of the Periodic Table are more electronegative than those below them and the electronegativity increases from left to right across any row. Fluorine is the most electronegative, oxygen and chlorine less so, and nitrogen, sulphur, and bromine less again. Hydrogen, carbon and iodine have almost identical electronegativities. The metals have low electronegativities and can be called electropositive.

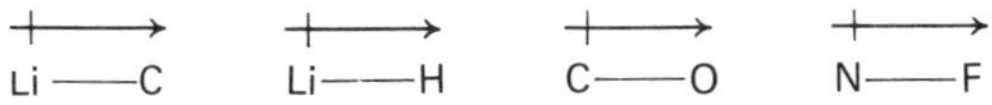

Examples of polar bonds

Examples of bonds of low or zero polarity

This means that in a molecule such as methyllithium, $CH_3-Li$, which is a reactive but ordinary covalent compound, the electron density at the carbon atom is higher than it is at the carbon atom of methane, whereas in chloromethane, $Cl-CH_3$, the electron density at carbon is lower than in methane.

What will happen if there are several polar bonds, as in water or chloroform (trichloromethane)? The problem is like that of two horses pulling equally on a load but in different directions. The load will move in a direction between the horses.

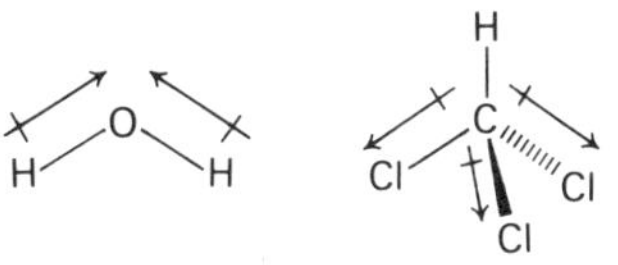

Bond dipoles

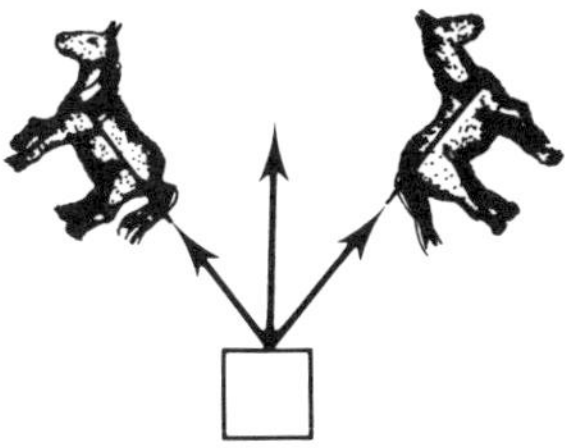

The polarity of the whole molecule will be the vector sum of the bond polarities, that is, the sum allowing for the fact that the bond polarities cancel one another out in certain directions and reinforce one another along one direction only.

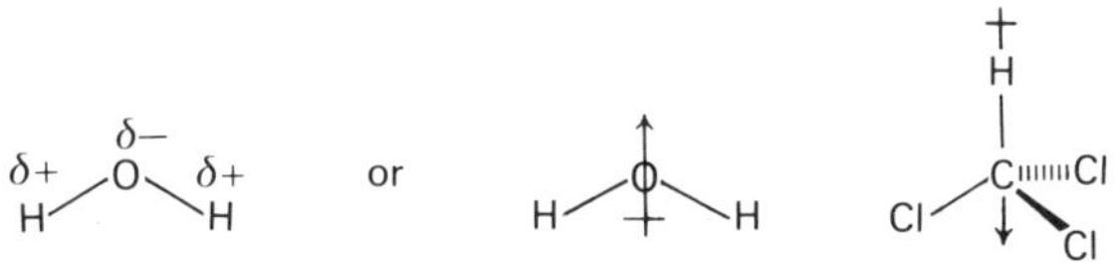

Resultant molecular dipoles

Chloroform and water are therefore polar molecules. Because of this, there is an electrostatic attraction between one molecule of chloroform and another—a dipole–dipole attraction.

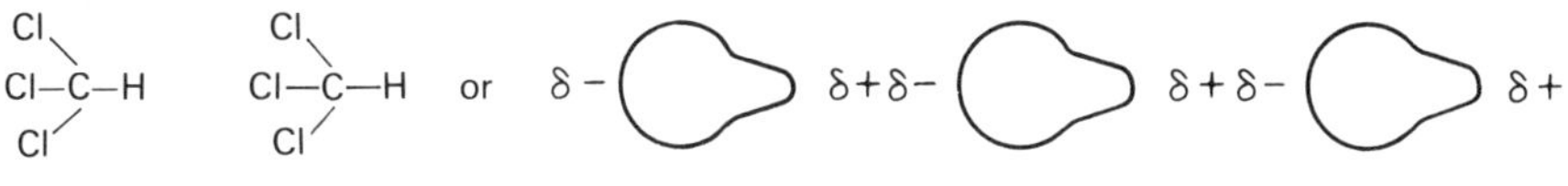

In carbon tetrachloride, although all the bonds are polar there is zero net dipole since the chlorine atoms are arranged symmetrically round the carbon atom and the polarities of the bonds cancel one another out. Carbon tetrachloride, like methane and bromine, Br—Br, is a totally non-polar molecule:

$$
\begin{array}{c}
\mathrm{Cl} \\
| \\
\mathrm{C}\text{\tiny\!\!\!\!\!\!IIIII}\mathrm{Cl} \\
\diagup\quad\diagdown \\
\mathrm{Cl}\quad\ \mathrm{Cl}
\end{array}
$$

Polar molecules such as water will orientate themselves in an electric field. If this field is due to an ion, they will arrange themselves radially, forming a three-dimensional cluster round the ion:

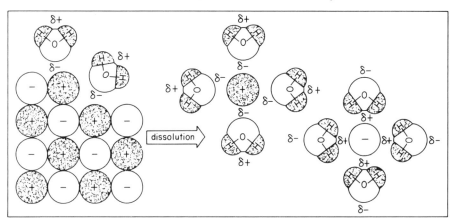

The attraction between the water molecules and the ions may be of comparable magnitude to the dipole–dipole attraction between water molecules and to the ion–ion attraction in the solid NaI. If this is the case, the ionic solid, e.g. sodium iodide, will dissolve in water as independent ions which are hydrated or *solvated*.

Polar solutes such as propan-2-one (acetone) can also be solvated, the molecules of water being orientated one way at one end and the other way at the other end. Propan-2-one is completely soluble in water.

# HYDROGEN BONDS

Molecules containing F—H, O—H, N—H, or S—H bonds are all polar but the strong attraction observed between such molecules is a special kind of dipole–dipole attraction. Molecules of water or alcohols tend to orient themselves so that the small $\delta+$ hydrogen atoms lie between the $\delta-$ oxygen atoms and form an electrostatic sandwich. This electrostatic attraction is called *hydrogen bonding*.

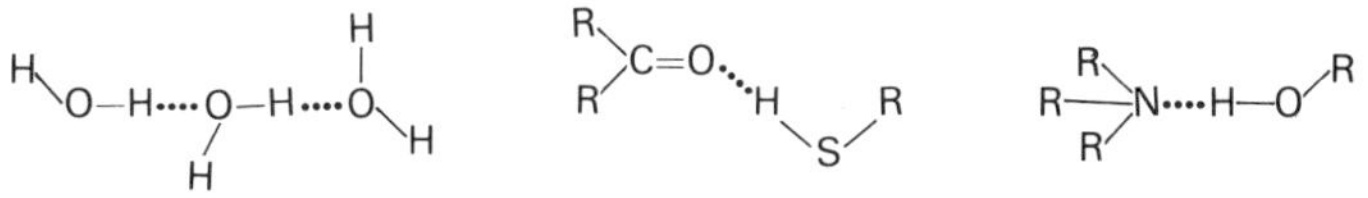

The hydrogen bond can be represented by a dotted line. The bond strengths of hydrogen bonds can be from 5 to 200 kJ mol$^{-1}$ while those of C—H and C—C covalent bonds are about 410 and 350 kJ mol$^{-1}$, respectively.

The presence of hydrogen bonding can have marked effects on solubility, boiling points and conformations and on adsorption on surfaces, including enzymes, and on the degree of dissociation of O—H groups. Without the hydrogen bond the sea might evaporate and all forms of life would cease since the code-book of inherited structural plans, contained in the deoxyribonucleic acid (DNA) double helix, could not be replicated. This is because in the synthesis of a new molecule of a nucleic acid by linking the appropriate small molecules, the latter are arranged in the correct order by lining them up along an existing nucleic acid molecule. The unconnected units are held to the 'template' molecule by hydrogen bonds.

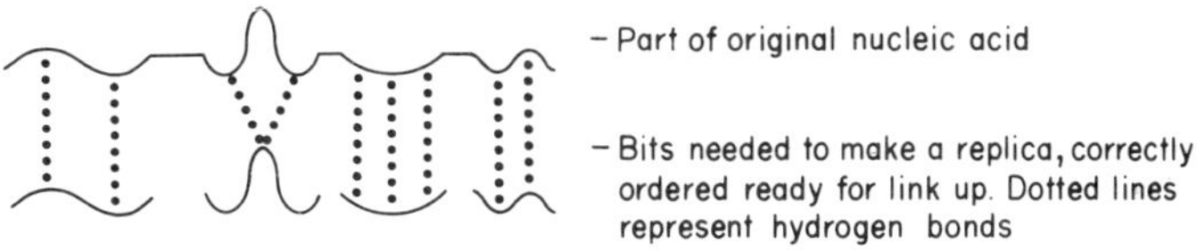

– Part of original nucleic acid

– Bits needed to make a replica, correctly ordered ready for link up. Dotted lines represent hydrogen bonds

Diagram of nucleic acid reproduction

*Problem 5–1. Suggest why the removal of a proton from the carboxyl group of the ion A is much easier than from its* cis *isomer B.*

*A*

*B*

*Problem 5–2. Suggest why 2-fluoroethanol adopts preferentially the conformation C rather than D.*

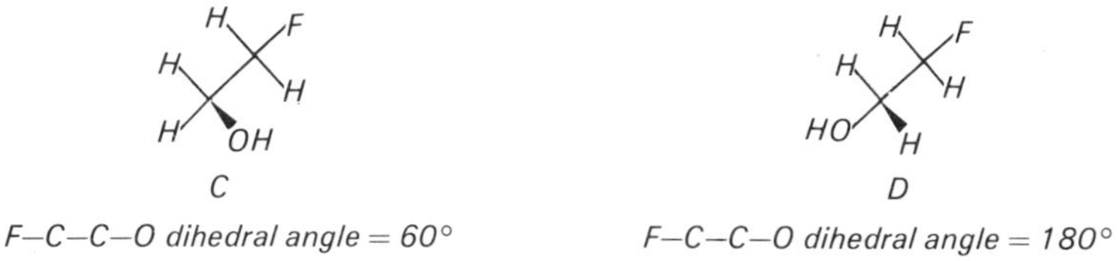

*C*

*D*

*F—C—C—O dihedral angle* $= 60°$

*F—C—C—O dihedral angle* $= 180°$

## ELECTRONEGATIVITY AND REACTION PATHWAYS

We have seen that when atoms of different electronegativity are bonded to one another, the electrons of the bond are not equally shared but are pulled towards the more electronegative atom, making the bond polar. If the electrons were to become associated solely with one atom, the bond would be broken and two independent fragments would be formed. If the original molecule had no net charge the fragments will be charged ions. Breaking the molecule up to give these oppositely charged ions requires the input of energy, but this can be compensated for by a reduction in energy due to attraction between the ions and polar solvent molecules (see NaI above). Ionization of molecules is therefore likely only in polar solvents.

X–Y may, in a suitable solvent, give $\overline{X}^{\ominus} + Y^{\oplus}$ or $^{\oplus}X + \overline{Y}^{\ominus}$. The direction in which the electrons move depends on the electronegativities of X and Y and also on the relative stability of the pairs of solvated ions. HCl gas is a polar molecule. When it dissolves in water, ionization to $H^{\oplus}$ (solvated) and $Cl^{\ominus}$ (solvated) occurs. $H^{\ominus}$ and $Cl^{\oplus}$ are not formed, one of the reasons being that chlorine is more electronegative than hydrogen.

Similarly, when 2-chloro-2-methylpropane is dissolved in a suitable solvent, ionization occurs to give a chloride ion and a carbonium ion, partly because chlorine is more electronegative than carbon:

$$\underset{\underset{CH_3}{|}}{\overset{\overset{CH_3}{|}}{|\overline{Cl}|\text{-}C\text{-}CH_3}} \rightarrow |\overline{Cl}|^{\ominus} + \underset{CH_3}{\overset{CH_3}{}}\overset{\oplus}{C}_{CH_3} \quad \left( not \ |\overline{Cl}|^{\oplus} + CH_3\overset{\overline{C}^{\ominus}}{\underset{CH_3}{|}}CH_3 \right)$$

The relative electronegativity of the atoms concerned is thus one of the factors (there can be others) which determine the path taken by bond-breaking reactions.

The path taken by bond-forming reactions can also be decided by electronegativity considerations. The addition of a hydride ion to a ketone, say propan-2-one (acetone), could give either an alkoxide ion by formation of a new H–C bond, or a carbanion by formation of a new H–O bond.

$$\underset{CH_3}{\overset{CH_3}{}}C=\overline{\underline{O}} + \overline{H}^{\ominus} \rightarrow \underset{CH_3}{\overset{CH_3}{}}C\overset{H}{\underset{\overline{O}^{\ominus}}{}}$$

An alkoxide ion

$$\underset{CH_3}{\overset{CH_3}{}}C=\overline{\underline{O}} + \overline{H}^{\oplus} \rightarrow \underset{CH_3}{\overset{CH_3}{}}\overset{\ominus}{C}\text{-}\overline{\underline{O}}\text{-}H$$

A carbanion

The negative charge is on a more electronegative atom in the alkoxide ion than in the carbanion. This factor favours the formation of the alkoxide ion and is one of the factors which makes the alkoxide ion the exclusive product of this addition. (Other examples are discussed in Chapter 15.)

# ACIDITY AND BASICITY

The behaviour of different molecules in the same type of reaction may also be related to the electronegativity of the atoms involved.

We shall examine the ionization of X—H bonds to give $X^{\ominus}$ and a proton. An *acid* has been defined as any substance which is a source of protons, or more exactly which can donate protons to other molecules such as the solvent:

$$X-H \rightarrow X^{\ominus} + H^{\oplus} \quad \text{or} \quad X-H + S \rightarrow X^{\ominus} + HS^{\oplus}$$

We define as a *base* any substance which can accept a proton, that is form a bond to a proton or capture one from another molecule:

$$X^{\ominus} + H^{\oplus} \rightarrow X-H \quad \text{or} \quad X^{\ominus} + HS \rightarrow X-H + S^{\ominus}$$

Aqueous HCl is sour and corrosive and we say it is *acidic*. Water is not. Yet both can be described as *acids* since both can act as proton donors.

Solutions of sodium hydroxide are soapy and corrosive. We say they are *alkaline*. Water is not. Yet both can be described as *bases* since both can act as proton acceptors.

$$H_2O + H^{\oplus} \rightarrow H_3O^{\oplus}$$
$$HO^{\ominus} + H^{\oplus} \rightarrow H_2O$$

Notice that a substance such as water can behave as either an acid or a base.

Let us now compare the acidity of the series of molecules:

$$F-H \qquad HO-H \qquad H_2N-H \qquad H_3C-H \qquad H-H$$

Hydrofluoric acid is a fairly strong acid. When dissolved in water it ionizes:

$$|\overline{F}-H \rightleftharpoons |\overline{F}|^{\ominus} + H^{\oplus}$$

and dissociation is virtually complete, that is nearly all of the molecules dissociate. The equilibrium constant is large. Water, on the other hand, dissociates to only a slight extent:

$$H\overline{\underline{O}}-H \rightleftharpoons H-\overline{\underline{O}}|^{\ominus} + H^{\oplus}$$

Ammonia dissociates to an even smaller extent, and methane and hydrogen do not dissociate at all, even in polar solvents.

The greater the electronegativity of the atom which captures the bonding electrons during dissociation, the greater is the degree of dissociation—at least in this series.

Let us now compare the strengths of bases. Anions are better captors of protons than neutral molecules because of the electrostatic attraction between negative anion and positive proton. Thus, hydroxide ion is a stronger base than water and the amide ion, $^{\ominus}NH_2$, is stronger than ammonia. However, ammonia is a stronger base than water and among the anions the amide ion is stronger than hydroxide ion and the carbanion $^{\ominus}CH_3$ is even stronger. We have, in fact, a series of anions of increasing basicity:

$$F^{\ominus} \qquad HO^{\ominus} \quad H_2N^{\ominus} \qquad H_3C^{\ominus}$$

*least basic* *most basic*

It appears that, in this series, the tendency to donate lone-pair electrons to form bonds increases as the electronegativity of the donor atom decreases. The atoms which have the strongest pull on bonding electrons are also least able to share any lone-pair electrons they may have.

It is interesting to note that whereas HF is a fairly strong acid $F^{\ominus}$ is a fairly weak base, and whereas $CH_4$ is a very weak acid $CH_3{}^{\ominus}$ is a very strong base. This should not be surprising. It is true of any acid and base which are related by the equilibrium

$$X-H \rightleftharpoons X^{\ominus} + H^{\oplus} \quad \text{or} \quad {}^{\oplus}X-H \rightleftharpoons X + H^{\oplus}$$

and is a direct consequence of the interrelationship of the acid and the base.

| | | | |
|---|---|---|---|
| F—H | $F^{\ominus}$ | $+ H^{\oplus}$ | The degree of dissociation is large. The equilibrium constant is large. |
| Strong acid | Weak base | | |
| $H_2\overset{\oplus}{O}-H$ | $H_2O$ | $+ H^{\oplus}$ | |
| Strong acid | Weak base | | |
| $H_3\overset{\oplus}{N}-H$ | $H_3N$ | $+ H^{\oplus}$ | The equilibrium constant is not far from 1. |
| Moderate acid | Moderate base | | |
| HO—H | $H-O^{\ominus}$ | $+ H^{\oplus}$ | |
| Weak acid | Strong base | | |
| $H_3C-H$ | $H_3C^{\ominus}$ | $+ H^{\oplus}$ | The degree of dissociation is very small. The equilibrium constant is very small. |
| Very weak acid | Very strong base | | |

The relative acidities and basicities are best expressed by the numerical values of the equilibrium constants. We shall return to this theme in Chapter 16. Since the ions or molecules may be solvated the exact values depend on the solvent.

What is clear from the discussion above is that electronegativity values are a useful guide in comparing both the basicities and the acidities of compounds of elements in any one row of the Periodic Table.

*Problem 5–3. If $H-O-CH_2-H$ (methanol) were to lose a proton to a base, which anion would be formed preferentially?*

*Problem 5–4. If $CH_3NH_3{}^{\oplus}$ is added to water what proton transfer process might occur? Will the equilibrium constant for this process be large or small?*

## SUMMARY

We have seen now that the polarity of molecules is a major factor in determining their properties—in particular their interaction with other molecules during solvation, adsorption, evaporation, etc. The polarity of the molecule can be viewed as the vector sum of the polarities of bonds. The polarities of bonds can be important in deciding interactions between groups within a molecule. The polarities of bonds can be summarized by a table of relative electronegativity values which quantify the attraction that the atoms of a bond have for the electrons of the bond. These electronegativity values can also be used to understand the course taken in certain reactions where there is a choice of path, and to understand the relative electron-donating abilities of atoms. Other effects can be more important than electronegativity in some cases. One of these other effects is discussed in the next chapter.

## ANSWERS TO PROBLEMS IN CHAPTER 5

*Problem 5–1.* Extra energy is required for the removal of the proton from the *cis* isomer, which is stabilized by intramolecular hydrogen bonding:

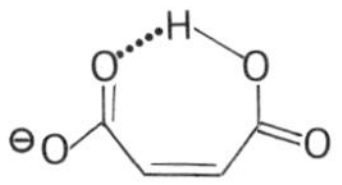

*Problem 5–2.* Intramolecular hydrogen bonding is possible when the molecule is in the conformation with dihedral angle 60° but not in the other.

*Problem 5–3.* $^{\ominus}|\underline{O}-CH_2-H$ rather than $H-\overline{\underline{O}}-\overset{\ominus}{C}H_2$ since the former is more stable. One reason for this is that the negative charge is then on a more electronegative atom.

*Problem 5–4.* The water could accept a proton from the $H_3C-\overset{\oplus}{N}H_3$. The equilibrium

$$CH_3\overset{\oplus}{N}H_3 + H_2O \rightleftharpoons CH_3NH_2 + H_3O^{\oplus}$$

will be set up and the equilibrium constant will be small, that is the protons will mostly be attached to the amine rather than the water, since amines are more basic than water. A positive charge on nitrogen is preferred to a positive charge on oxygen since the former is less electronegative.

# Chapter 6

## Delocalization

Chapter 2 discussed how to work out the number of valency electrons present in a molecule, how to allocate these as bonding pairs or lone pairs so as to give all of the atoms probable arrangements of valency electrons, and how to work out which atoms would then be charged. The result of this procedure was one, or sometimes more than one, provisional or possible electronic structure for the molecule. In many cases one electronic structure is clearly more satisfactory than the others and this one often provides an adequate basis for discussion of the shape or the reactions of the molecule.

For example, the ion $(CN)^{\ominus}$ must have $4 + 5 + 1 = 10$ valency electrons, i.e. five pairs, and if these are arranged as $^{\ominus}\overline{C}\equiv N$, which requires that the negative charge be on the carbon atom, we arrive at a model or representation of the electronic arrangement for the ion which adequately accounts for its physical and chemical properties. The alternative, $\overline{C}=\underline{\overline{N}}^{\ominus}$, is probably a poorer representation of the actual electron distribution since both atoms do not in this case have complete octets.

However, if we repeat this process for the nitrite ion, $(ONO)^{\ominus}$, we arrive at two different structures of equal plausibility:

$$^{\ominus}|\overline{\underline{O}}-\overline{N}=\overline{\underline{O}} \quad \text{and} \quad \overline{\underline{O}}=\overline{N}-\overline{\underline{O}}|^{\ominus}$$

One oxygen atom carries the negative charge in the first case, and the other oxygen atom carries it in the second case. We must now ask whether in the nitrite ion the negative charge *is* all at one end, or is the real electron distribution such that neither of these pictures is very near the truth.

Similarly, electron distributions for the ion $(H_2COH)^{\oplus}$ might be drawn such as

$$\underset{H}{\overset{H}{>}}\overset{\oplus}{C}-\overset{-}{\underset{H}{\overset{\backslash}{O}}} \quad \text{or} \quad \underset{H}{\overset{H}{>}}C=\overset{\oplus}{\underset{H}{\overset{\backslash}{O}}}$$

Again, we must ask what the actual distribution of the charge is. Is the bond between the C and the O a double bond or a single bond, or is it perhaps something in between?

The evidence, from structural studies and from the courses and energetics of their reactions, is that the actual electron distribution in a molecule such as $(H_2COH)^{\oplus}$ cannot be described by any one simple picture of the above type, because the real electron density between the C and the O is somewhere intermediate between what is normal for a true single bond and what is normal for a true double bond. In other words, the C—O bond present in this ion has a character intermediate between that of a double bond and a single bond. The actual non-bonding electron density on the oxygen is somewhere between 2 and 4 electrons and the positive charge is carried partly by the C and partly by the O.

The problem is therefore not so much to establish what the real electron distribution is, as we can determine that by experiment, but rather the way in which this molecule is to be represented. How should we represent the partial double bond character of the C—O bond and the sharing of the positive charge? There is no entirely satisfactory solution.

Since we decided earlier to describe all of the molecules by using only lines and full charges, we are forced to say that the real electron distribution is something like

$$\begin{array}{c} H \\ \diagdown \overset{\oplus}{C} - \overset{\ominus}{O} \diagdown \\ H \diagup \qquad \diagdown H \end{array}$$

and something like

$$\begin{array}{c} H \\ \diagdown C = \overset{\oplus}{O} \diagdown \\ H \diagup \qquad \diagdown H \end{array}$$

and is intermediate between these. We say that the actual structure is a *hybrid* of these two simple structures. We draw the following to represent the actual electron arrangement:

$$\begin{array}{c} H \\ \diagdown \overset{\oplus}{C} - \overset{\ominus}{O} \diagdown \\ H \diagup \qquad \diagdown H \end{array} \quad \leftrightarrow \quad \begin{array}{c} H \\ \diagdown C = \overset{\oplus}{O} \diagdown \\ H \diagup \qquad \diagdown H \end{array}$$

The whole drawing must be regarded as a unity. Note the symbol $\leftrightarrow$, which is used for this and no other purpose.

We might try to represent the actual molecule by

$$\begin{array}{c} H \\ \diagdown \overset{\oplus}{C \cdots O} \diagdown \\ H \diagup \qquad \diagdown H \end{array}$$

or some other drawing of the partial double bond character of the C—O bond, but this has the disadvantage of introducing new symbols. Alternatively, we could simply represent the molecule as

$$\begin{array}{c} H \\ \diagdown \overset{\oplus}{C} - \overset{\ominus}{O} \diagdown \\ H \diagup \qquad \diagdown H \end{array}$$

(or the other structure) and *remember* that this is a rather poor approximation of the real situation.

There are many molecules whose bonds are not simple 2-electron, 4-electron, or 6-electron bonds, and this problem of drawing such molecules will arise frequently in the later chapters. Examples of two important structures of this type, benzene and carboxylic acids, are discussed in more detail in Chapters 14 and 16, respectively.

Experiments indicate that in the nitrite ion the charges on the oxygen atoms are identical and the two N–O bonds are identical, each being half way between a single and a double bond. The nitrite ion can therefore be represented by

$$\overset{\ominus}{O}\diagdown\overset{\bar{N}}{}\diagup\diagdown O \quad\leftrightarrow\quad O\diagup\diagup\overset{\bar{N}}{}\diagdown\diagdown O^{\ominus} \qquad \text{or by} \qquad O\cdots\overset{\cdots}{N}\cdots O$$

Ketones are usually drawn as $>C=\bar{O}$. The arrangement $>\overset{\oplus}{C}-\bar{O}|^{\ominus}$ might be drawn but the experimental evidence is that $>C=\bar{O}$ is a very good representation of the actual electron arrangement. Thioketones, however, have properties which show that $>C=\bar{S}$ is not a very good representation. It would be nearer the truth to depict them as

$$>C=\bar{S} \quad\leftrightarrow\quad >\overset{\oplus}{C}-\bar{S}|^{\ominus}$$

In molecules such as $(H_2COH)^{\oplus}$, $NO_2^{\ominus}$, $(CH_2CHCH_2)^{\oplus}$, $(HCO_2)^{\ominus}$, $RN_2^{\oplus}$, $NCS^{\ominus}$, and $CO_3^{2-}$, some of the electrons are not localized entirely on one atom or between two atoms but are spread over two or three atoms. They are *delocalized* electrons. Some chemists use the word *resonance* to describe the situation in a molecule which has delocalized electrons. Remember that for all of these molecules there is no doubt about the structure and no doubt about the actual electron distribution—the difficulty is only one of drawing them.

*Problem 6–1. Draw representations using the symbol $\leftrightarrow$ for all the substances listed in the previous paragraph. The skeletons are as follows:*

$$C\diagup^{C}\diagdown C \qquad H-C\diagdown\diagup^{O}_{O} \qquad R-N-N \qquad N-C-S \qquad \overset{O}{\diagdown}C-O\diagup^{O}$$

## PROPERTIES OF MOLECULES WITH DELOCALIZED ELECTRONS

What will be the effects of the delocalization of electrons in a molecule? One important effect is on the *conformation*. Rotation round single bonds in molecules is usually easy, whereas rotation round double bonds is usually impossibly difficult. Rotation round a bond of intermediate character should probably be rather difficult. Molecules such as ethene, with double bonds, are entirely planar. They can have either of two forms (*cis* and *trans* when these are distinguishable) which cannot be interconverted by rotation.

A suitable example of a molecule with delocalized electrons is an amide. Amides were represented in Chapter 3 by the simple formula $RCONH_2$. The evidence is that this simple picture is not very accurate and they are better represented by a composite picture:

$$\overset{O}{\diagdown}C-\bar{N}\diagup^{H}_{H} \quad\leftrightarrow\quad \overset{O^{\ominus}}{\diagdown}C=\overset{\oplus}{N}\diagup^{H}_{H}$$

Amides are polar molecules with a partial negative charge on the oxygen atom and a partial positive charge on the nitrogen atom. The C—N bond has the character of a double bond to a considerable extent. Because of this, the whole amide group is planar and rotation about the C—N bond is difficult.

A second important effect of delocalization is that molecules having delocalized electrons are more *stable* than similar molecules which do not. For example, the delocalized ion

$$CH_3-CH=CH-\overline{\underline{O}}|^{\ominus} \quad \leftrightarrow \quad CH_3-\overset{\ominus}{C}H-CH=\overline{\underline{O}}$$

is more stable than either of the ions below, which are both isomers of it:

$$CH_2=CH-CH_2-\overline{\underline{O}}|^{\ominus} \quad \text{or} \quad ^{\ominus}\overline{C}H_2-CH_2-CH=\overline{\underline{O}}$$

The ion $CH_3-CH=CH-\overset{\oplus}{C}H_2 \leftrightarrow CH_3-\overset{\oplus}{C}H-CH=CH_2$ is more stable than its isomer $^{\oplus}CH_2-CH_2-CH=CH_2$, which is not delocalized; and the ion $^{\oplus}CH_2-\overline{\underline{O}}H$ $\leftrightarrow CH_2=^{\oplus}\overline{O}H$ is much more stable than the ion $^{\oplus}CH_3$. Consideration of the electronegativities of H and O would lead to the opposite conclusion, since one might expect that an ion with a partial positive charge on the oxygen atom would be less stable than one with a positive charge on the carbon atom, or that a cation $^{\oplus}CH_3$ would be made less stable by replacing one of the hydrogens by an oxygen group as in $^{\oplus}CH_2OH$. In fact, however, the ion is not $^{\oplus}CH_2OH$; it is a delocalized ion and the stabilization due to delocalization of electrons is greater than the destabilization due to having a partial positive charge on an electronegative atom. In molecules with delocalized electrons the electronegativity effects discussed in Chapter 5 are usually swamped by the effects of this delocalization.

*Problem 6–2. Treatment of*

$$CH_3-\underset{\underset{CH_3}{|}}{\overset{\overset{H}{|}}{C}}-C\overset{H}{\underset{O}{\diagdown}}$$

*with base results in removal of a proton to form an anion. The reaction is under thermodynamic control, i.e. the most stable anion is the one formed. Which proton do you expect to be removed?*

## ANSWERS TO PROBLEMS IN CHAPTER 6

*Problem 6–1.*

$$\overset{H}{\underset{H}{\diagup}}C-\overline{\underline{O}}\overset{\oplus}{\underset{H}{\diagdown}} \quad \leftrightarrow \quad \overset{H}{\underset{H}{\diagup}}C=O\overset{\oplus}{\underset{H}{\diagdown}}$$

$$^{\ominus}\underset{\diagup}{\overline{O}}\diagup\overline{N}\diagdown\underset{\diagdown}{O} \quad \leftrightarrow \quad \underset{\diagup}{O}\diagup\overline{N}\diagdown\underset{\diagdown}{O}^{\ominus}$$

$$\underset{H_2C}{\diagup}\overset{\overset{H}{|}}{\underset{\diagdown}{C}}\overset{\oplus}{\underset{CH_2}{\diagdown}} \quad \leftrightarrow \quad H_2C\overset{\oplus}{\diagup}\overset{\overset{H}{|}}{\underset{\diagdown}{C}}\diagdown CH_2$$

The end carbon atoms each carry half a positive charge.

60

$$H-C\overset{O}{\underset{O^{\ominus}}{\diagup}} \quad \longleftrightarrow \quad H-C\overset{O^{\ominus}}{\underset{O}{\diagup}}$$

The oxygen atoms are equivalent and each carries half a negative charge.

$$R-\overset{\oplus}{N}\equiv\bar{N} \quad \longleftrightarrow \quad R-\bar{N}=\bar{N}^{\oplus}$$

The nitrogen atoms each carry part of the positive charge.

$$^{\ominus}\bar{N}=C=\bar{S} \quad \longleftrightarrow \quad \bar{N}\equiv C-\underline{S}|^{\ominus}$$

The nitrogen and the sulphur atoms each carry part of the negative charge.

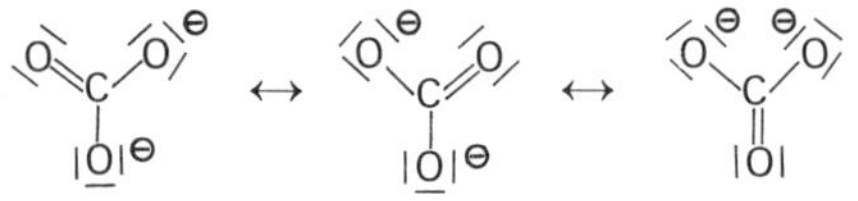

$$\overset{O}{\diagup}C\overset{O^{\ominus}}{\diagdown} \quad \longleftrightarrow \quad \overset{O^{\ominus}}{\diagdown}C\overset{O}{\diagup} \quad \longleftrightarrow \quad \overset{O^{\ominus}\ O^{\ominus}}{\diagdown}C\overset{}{\diagup}$$

Three components are needed to give an adequate picture. All three oxygen atoms are equivalent and each carries two-thirds of a negative charge.

*Problem 6–2.* The proton on the carbon next to the carbonyl group is removed to form the delocalized ion

which is more stable than either of the alternative ions.

# Chapter 7

## Reaction Types, Mechanistic Drawings, and Reaction Product Selection

### TYPES OF REACTION

Now that we can draw structures for molecules and make deductions about their shapes and their polarities (and hence many of their physical properties), we can turn to their reactions.

In a reaction, bonds are broken and bonds are formed. The sharing and distribution of electrons among the nuclei changes. The details of what happens to molecules when they react, and in particular what happens to the electrons, define the *mechanism* of the reaction.

We can begin our study of mechanisms by looking at and distinguishing between the kinds of electron redistributions which can occur in simple one-step reactions. A list of examples of such reactions is given below. We need not bother here with the nature of the groups R or the conditions for these reactions. They have been grouped according to what happens to the electrons. Remember that the nature of the nuclei of the atoms involved is not important. The same *types* of electron reorganizations could occur with other nuclei. We can distinguish four types, which are discussed more fully below.

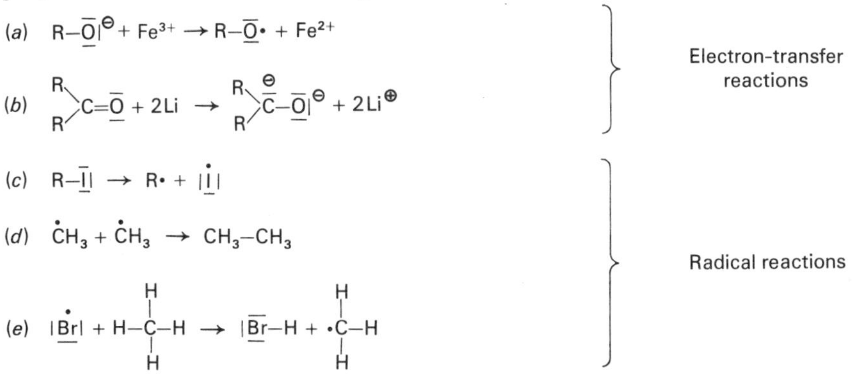

(a) $R-\overline{\underline{O}}|^{\ominus} + Fe^{3+} \rightarrow R-\overline{\underline{O}} \cdot + Fe^{2+}$

Electron-transfer reactions

(b) $\begin{array}{c} R \\ \diagdown \\ R \diagup \end{array} C=\overline{\underline{O}} + 2Li \rightarrow \begin{array}{c} R \\ \diagdown \\ R \diagup \end{array} \overset{\ominus}{C}-\overline{\underline{O}}|^{\ominus} + 2Li^{\oplus}$

(c) $R-\overline{\underline{I}}| \rightarrow R\cdot + |\underline{\dot{I}}|$

(d) $\dot{C}H_3 + \dot{C}H_3 \rightarrow CH_3-CH_3$

Radical reactions

(e) $|\dot{Br}| + H-\overset{\overset{\displaystyle H}{|}}{\underset{\underset{\displaystyle H}{|}}{C}}-H \rightarrow |\overline{Br}-H + \cdot\overset{\overset{\displaystyle H}{|}}{\underset{\underset{\displaystyle H}{|}}{C}}-H$

61

62

(f)  $H-\overline{O}-H \rightarrow H^{\oplus} + {}^{\ominus}|\overline{O}-H$

(g)  ${}^{\ominus}\overline{C}\equiv N + {}^{H}_{H}{>}C=\overline{O} \rightarrow \overline{N}\equiv C-\overset{\overset{\displaystyle H}{|}}{\underset{\underset{\displaystyle H}{|}}{C}}-\overline{O}|^{\ominus}$   } Ionic reactions

(h)  $H-CH_2-CH_2-\overset{\oplus}{\underset{\underset{\displaystyle H}{|}}{O}}-H \rightarrow H^{\oplus} + CH_2{=}CH_2 + H-\overline{O}-H$

(i)    } Pericyclic reactions

(j)

### Electron-transfer reactions

In reaction (a), the alkoxide ion, $R-\overline{O}|^{\ominus}$, loses entirely one electron, which is picked up by the iron(III) ion. The iron(III) ion is reduced to an iron(II) ion and the alkoxide ion is oxidized to a radical (see p. 13). There is total transfer of one electron from one reagent to the other.

(a)  $R-\overline{O}|^{\ominus} + Fe^{3+} \rightarrow R-\overline{O}\cdot + Fe^{2+}$

(b)  ${}^{R}_{R}{>}C=\overline{O} + 2Li \rightarrow {}^{R}_{R}{>}\overset{\ominus}{\underset{}{C}}-\overline{O}|^{\ominus} + 2Li^{\oplus}$

In reaction (b), there is *total* transfer of two electrons from the lithium metal to the ketone to give two lithium ions and a doubly negative organic ion. Reactions (a) and (b) are examples of one type of reaction which we can call *electron-transfer reactions*. They are *redox* reactions but not all reduction/oxidation reactions proceed by this type of mechanism.

### Radical reactions

Reactions (c), (d), and (e) are of a different type called *radical reactions* since, in each, some of the reagents or products are radicals but in none is there total transfer of electrons from one reacting species to the other. In reaction (c) the iodoalkane molecule breaks up or dissociates to give an alkyl radical and an iodine atom.

(c)  $R-\overline{I}| \rightarrow R\cdot + |\overline{\underline{I}}|$

(d)  $\overset{\bullet}{C}H_3 + \overset{\bullet}{C}H_3 \rightarrow CH_3-CH_3$

(e)  $|\overset{\bullet}{\underline{B}r}| + H-\overset{\overset{\displaystyle H}{|}}{\underset{\underset{\displaystyle H}{|}}{C}}-H \rightarrow |\overline{\underline{B}}r-H + \bullet\overset{\overset{\displaystyle H}{|}}{\underset{\underset{\displaystyle H}{|}}{C}}-H$

The C—I bond is broken in such a way that one electron of the bonding pair ends up with the carbon and the other with the iodine. Such a bond breaking is described as *homolysis* of the bond. Homolysis is derived from two Greek words, *homo*—same and *lysis*—splitting, so the word means splitting into similar things, in this case two radicals. The bonded atoms receive one electron each. You may have met the words homogeneous, homogenized, or homosexual. They all contain the prefix meaning 'of the same kind'.

In reaction (d) a radical combination occurs. Two molecules, each with an odd number of electrons, combine so that all electrons become paired. In reaction (e) a radical reacts with a molecule with all-paired electrons to give a new radical and a new all-paired molecule. Here a hydrogen atom is transferred from carbon to bromine. The H—C bond is broken, one electron remaining with the carbon, and the hydrogen nucleus and the other electron combine with the bromine atom to form a molecule of HBr. All this happens at once. In all of these radical reactions unpairings or pairings of electrons occur. Bonding electrons become non-bonding (single) electrons or *vice versa*, but there is no total transfer of electrons from one atom to another.

## Ionic reactions

Reaction (f) is different again. Here a molecule is dissociating into fragments in such a way that the bonding pair of electrons stay paired. The H—O bond undergoes *heterolysis* (splitting into different kinds of things). Both electrons of the bonding pair end up on one of the atoms, in this case as a lone pair on the oxygen atom. The hydrogen loses a share in the bonding pair and becomes a cation.

(f)  $H-\bar{\underline{O}}-H \rightarrow H^{\oplus} + {}^{\ominus}|\bar{\underline{O}}-H$

(g)  ${}^{\ominus}\bar{C}{\equiv}\bar{N} + \overset{H}{\underset{H}{>}}C{=}\bar{\underline{O}} \rightarrow \bar{N}{\equiv}C-\overset{H}{\underset{H}{C}}-\underline{\bar{O}}|^{\ominus}$

(h)  $H-CH_2-CH_2-\overset{\oplus}{\underset{H}{O}}-H \rightarrow H^{\oplus} + CH_2{=}CH_2 + H-\bar{\underline{O}}-H$

Reaction (g) is similar. The lone pair on the carbon atom of the cyanide ion becomes shared between that carbon and the carbon atom of the formaldehyde. One of the bonds of the C=O double bond is broken and the bonding pair becomes a lone pair on the oxygen atom. One new bond has been formed and one old one broken. The electrons have remained in pairs. Shared electrons have become unshared (lone) pairs and *vice versa*. Again there is no total transfer of electrons. Reaction (h) also shows all of these characteristics.

Reactions (f), (g), and (h) are called *ionic reactions* since they necessarily involve ions as either reagents or products or both. In each case we can detect a direction of movement of electrons along the chain of atoms in the transition state.

## Pericyclic reactions

In reactions (*i*) and (*j*) we cannot detect a direction of movement of electrons. Here again the electrons stay paired. Bonds are broken and new ones formed and [in (*j*)] bonding electrons become non-bonding lone pairs. The reactions occur in one step via a cyclic transition state and a reorganization of the electrons occurs, but with no clear-cut movement of electrons in any one direction round the ring of atoms. Such reactions have been called electrocyclic or *pericyclic* reactions.

(*i*)

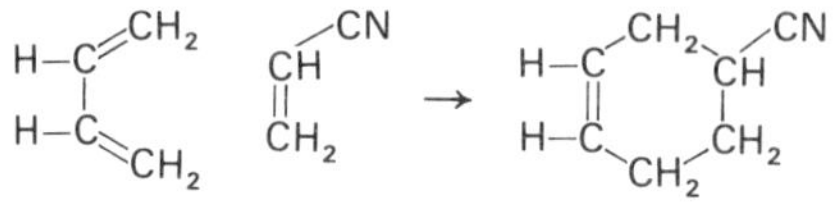

(*j*)

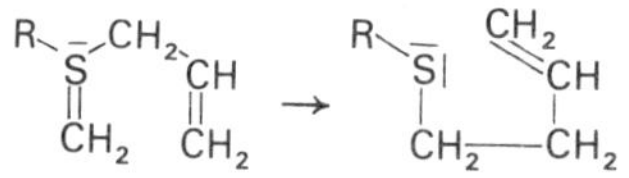

There are, therefore, four distinguishable types of reaction mechanisms—electron-transfer reactions, radical reactions, ionic reactions and pericyclic reactions. Examples of all of them can be readily found among chemical reactions from biochemical processes to petrochemical syntheses. Each of them is divisible into a few subtypes which we can usefully distinguish as we have done already for radical reactions and will do below for ionic reactions. The important conclusion is that all organic reaction steps belong to one or another of about a dozen basic mechanistic types or patterns. Organic reactions can therefore be classified simply (and learned more easily) by looking at their mechanisms.

## DRAWING IONIC MECHANISMS

This book is concerned largely with ionic reactions. In ionic reactions, electrons stay paired and move in a clearly detectable direction along a chain of atoms, in that an atom at one end loses a half share in a pair of electrons and an atom at the other end gains control over half an electron pair.

Molecules or ions which can accept electrons to form a bond are called *electrophiles* (electron seekers) or *Lewis acids* (after G. N. Lewis, who suggested this extension of the meaning of the word acid). It is clear that a proton can be electrophilic and so can formaldehyde, since the electrophilic carbon atom can accept electrons from a reagent to form a new bond with concurrent breakage of an existing bond.

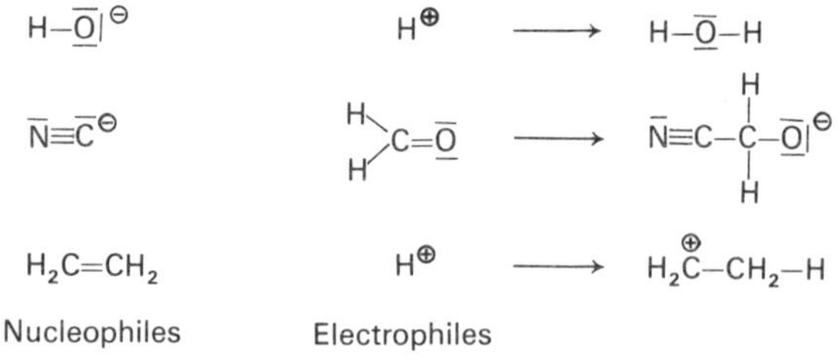

Nucleophiles          Electrophiles

Molecules or ions which can donate electrons to form a bond are called *nucleophiles* or *Lewis bases*. Hydroxide ion and cyanide ion can both behave as nucleophiles and both use lone-pair electrons to form new bonds. Ethene, $CH_2=CH_2$, can also behave as a nucleophile. In this case, the electrons which are used to form the new C—H bond were originally bonding electrons of the C=C double bond.

Many ionic reactions result from the interaction of a nucleophile and an electrophile.

In the above discussion, the mechanisms of the reactions have been described in words. It would be very useful to have a neat pictorial representation of the mechanism. Looking again at the equations (*f*), (*g*), and (*h*) above, we see that the positions of the electron pairs are known at the beginning and at the end of the reaction. The net effect of the reaction is that some of them have moved to adjacent positions, rather like matchsticks on the board in some childrens' game. To represent what has happened, we draw curved arrows on to the formulae of the reagent molecules, one for each pair of electrons that has moved, so that the tail of the arrow is where the electron pair is in the reagent, and the head of the arrow is where the electron pair ends up in the product. The mechanism of the dissociation of water could then be drawn as

$$\text{H}-\overline{\underline{\text{O}}}\text{H} \longrightarrow \text{H}-\overline{\underline{\text{O}}}|^{\ominus} + \text{H}^{\oplus}$$

Three of the electron pairs in the reagent (water) molecules are not involved in the reaction and remain in their original roles or positions at the end. One pair ceases to be shared between oxygen and hydrogen and becomes a lone pair on the oxygen; its altered role is indicated by the curved arrow.

The opposite reaction, the recombination of $H^{\oplus}$ and $^{\ominus}OH$, could be drawn as

$$\text{H}^{\oplus} \quad {}^{\ominus}|\overline{\underline{\text{O}}}-\text{H} \longrightarrow \text{H}-\overline{\underline{\text{O}}}-\text{H}$$

Note that, before drawing arrows, all electron pairs should be drawn in.

*Problem 7–1. Draw reactions (g) and (h) again and insert the correct arrows.*

If the electron arrangements of reagents and products in an ionic reaction are known, it is easy to insert arrows which summarize the change neatly.

It should also be easy to work out what the products would be if the reagents and the arrows describing the reaction course are given. Work out the structures of the products of the following transformations:

(*k*)   $\text{H}-\overset{\overset{\text{H}}{|}}{\underset{\underset{\text{H}}{|}}{\text{N}}}|$   $\text{H}^{\oplus}$   $\longrightarrow$

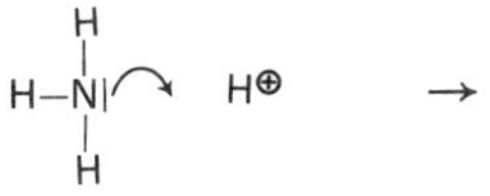

(*l*)   $\text{H}-\overline{\underline{\text{O}}}|^{\ominus}$   $\text{H}-\text{C}\equiv\text{N}$   $\longrightarrow$

(*m*)   $\text{H}-\overline{\underline{\text{O}}}|^{\ominus}$   $\text{CH}_3-\overline{\underline{\text{Br}}}|$   $\longrightarrow$

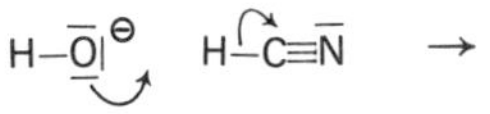

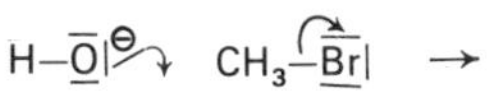

The full mechanistic equation for reactions (g) to (m) are given below.

(g) $\bar{N}\equiv\overset{+}{C}\ominus$ + $H_2C=\bar{O}$ → $\bar{N}\equiv C-\overset{H}{\underset{H}{C}}-\bar{O}|\ominus$

(h) $H-CH_2-CH_2-\overset{\oplus}{\underset{H}{O}}-H$ → $H^\oplus + CH_2=CH_2 + |\bar{O}-H$ with second H below O

(k) $H-\overset{H}{\underset{H}{N}}| \;\; H^\oplus$ → $H-\overset{H}{\underset{H}{N}}{}^\oplus-H$

(l) $H-\bar{O}|\ominus \;\; H-C\equiv N$ → $H-\bar{O}-H + \ominus\bar{C}\equiv\bar{N}$

(m) $H-\bar{O}|\ominus \;\; CH_3-\bar{Br}|$ → $H-\bar{O}-CH_3 + |\bar{Br}|\ominus$

In (k), (l) and (m), the new positions of the electron pairs follow directly from the positions of the arrows. If you did not correctly position the charges in the product molecules, look again at Chapter 2.

It is obvious from these examples and (if you think about it) necessary that the total net charge of the products is the same as that of the reactants. In reaction (f), water (net charge zero) gives a proton plus a hydroxide ion (net charge zero). In (k), ammonia plus a proton (net charge +1) gives ammonium ion (net charge +1).

In (l), it does not matter which lone pair of electrons on the oxygen atom is drawn as moving, as they are all indistinguishable. Nor does it matter how the arrows curl, provided that they start where a pair of electrons is and end where the pair goes. The mechanism could equally well have been drawn as follows

$$H-\bar{O}|\ominus \;\; H-C\equiv N \;\rightarrow\; H-\bar{O}-H + \ominus\bar{C}\equiv\bar{N}$$

When drawing mechanisms of ionic reactions by using arrows, usually no attempt is made to represent the correct stereochemistry of the reagents or products. This could be deduced from their electronic structures if needed (Chapter 4). Nor is any attempt made to represent the correct shape of the transition state, although this can sometimes be deduced from the shapes of reagent and product. By drawing reaction (g) as above, there is no implication that the N, C, C, and O atoms are collinear nor in (h) that the H, C, C, and O atoms are collinear. In fact they are not. We cannot explore this interesting subject here but you might like to think about the shapes of these transition states yourself.

Mechanistic drawings are really diagrams, and clumsy drawings such as

$$H-\bar{O}|\ominus + CH_3-C\!\!\begin{array}{c}O\\O-H\end{array} \;\rightarrow\; H-\bar{O}-H + CH_3-C\!\!\begin{array}{c}O\\O\,\ominus\end{array}$$

| Hydroxide ion | Acetic acid (ethanoic acid) | Water | Acetate ion (ethanoate ion) |

can be improved by redrawing the reagents:

$$H-\overline{\underline{O}}|^{\ominus} + H-\overline{\underline{O}}-\underset{\underset{|\underline{O}|}{\|}}{C}-CH_3 \rightarrow H-\overline{\underline{O}}-H + {}^{\ominus}|\overline{\underline{O}}-\underset{\underset{|\underline{O}|}{\|}}{C}-CH_3$$

This acetate ion reminds us of another problem. This ion is a delocalized ion, which would be more accurately drawn as

$$CH_3-C\overset{\nearrow O}{\underset{\searrow \underline{O}}{}}{}^{\ominus} \longleftrightarrow CH_3-C\overset{\nearrow O^{\ominus}}{\underset{\searrow \underline{O}}{}}$$

Acetate ion

When drawing the reactions of delocalized molecules, the most convenient approach is to represent them (simply but inaccurately) by only one electron picture.

Following this convention, the acetate ion produced in the reaction above was represented by one simple electron picture only. The situation is very slightly more complicated if the delocalized molecule is a reagent. Carbonate ion reacts with protons to give the hydrogencarbonate ion. We can draw the reaction starting with any one of the (three) pictures for the carbonate ion (see p. 60).

$$\overline{O}=C\overset{\nearrow O^{\ominus}}{\underset{\searrow O^{\ominus}}{}} H^{\oplus} \rightarrow \overline{O}=C\overset{\nearrow O^{\ominus}}{\underset{\searrow O-H}{}}$$

or

$$^{\ominus}|\overline{O}-C\overset{\nearrow O^{\ominus}}{\underset{\searrow O}{}} H^{\oplus} \rightarrow \overline{O}=C\overset{\nearrow O^{\ominus}}{\underset{\searrow O-H}{}}$$

Either of these representations is adequate. Note that one requires more arrows than the other. This is an accident resulting from the choice of which of the three (actually equivalent) oxygen atoms we draw as the proton acceptor.

The use of arrows to represent mechanisms is a very convenient shorthand notation, and it is easy to understand if it is done correctly. The beginner must draw in all the lone pairs of electrons, at least in the region of the molecule where changes are occurring, and must draw in one arrow for each electron pair which moves. The expert must continue to do the latter, but may omit the lone pairs, keeping the arrows in the same places as before, provided that he has mastered the interrelationship of the charge on an atom and electron distribution (Chapter 2). We shall take no short cuts until Chapter 17.

### Note

Several textbooks and some chemists use mechanistic arrows in a slipshod manner. Be systematic and *avoid* short cuts such as the following:

68

(*i*) the drawing of only the lone pair of electrons that moves when there are others on the same atom, e.g.

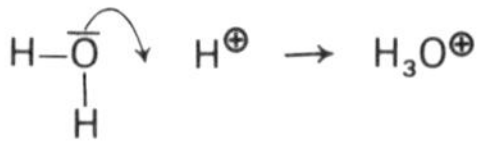

$$H\text{–}\overset{\frown}{\overline{O}} \quad H^{\oplus} \quad \rightarrow \quad H_3O^{\oplus}$$

which should be

$$H\text{–}\overline{O}| \quad H^{\oplus} \quad \rightarrow \quad H\text{–}\overset{\oplus}{O}\text{–}H$$

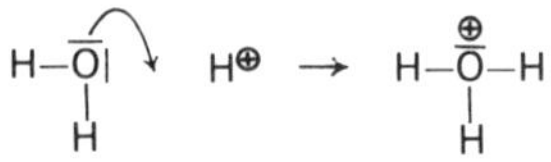

(*ii*) the drawing of arrows starting at minus signs as if it were the minus sign which moved or as if the minus represented a lone pair of electrons, e.g.

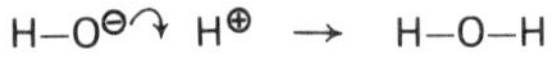

$$H\text{–}O^{\ominus} \quad H^{\oplus} \quad \rightarrow \quad H\text{–}O\text{–}H$$

which should be

$$H\text{–}\overline{O}|^{\ominus} \quad H^{\oplus} \quad \rightarrow \quad H\text{–}\overline{O}\text{–}H$$

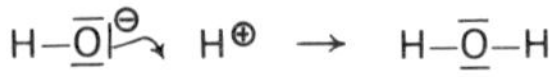

(*iii*) the omission from the mechanism of steps which must have occurred in view of the product written, e.g.

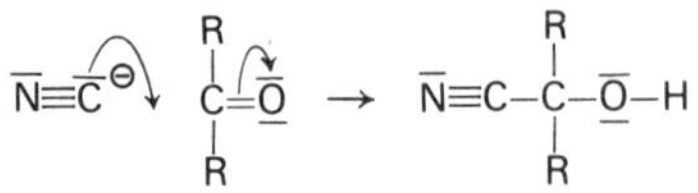

$$\overline{N}{\equiv}\overline{C}^{\ominus} \quad \overset{R}{\underset{R}{C}}{=}O \quad \rightarrow \quad \overline{N}{\equiv}C\text{–}\overset{R}{\underset{R}{C}}\text{–}\overline{O}\text{–}H$$

which should be

$$\overline{N}{\equiv}\overline{C}^{\ominus} \quad \overset{R}{\underset{R}{C}}{=}\underline{O} \quad \rightarrow \quad \overline{N}{\equiv}C\text{–}\overset{R}{\underset{R}{C}}\text{–}\overline{O}|^{\ominus}$$

$$NC\text{–}CR_2\text{–}\overline{O}|^{\ominus} \quad H^{\oplus} \quad \rightarrow \quad NC\text{–}CR_2\text{–}\overline{O}\text{–}H$$

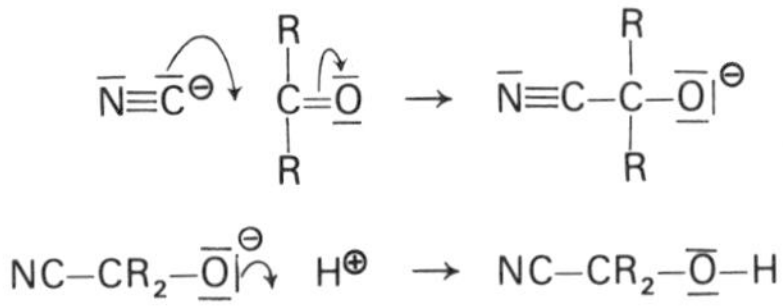

## PRODUCT SELECTION

Many organic reactions give several products. It is important to understand the factors which decide which product will be the major one. We shall consider a reagent R which could produce either product P or product Q. There are two distinct situations possible—

(*a*) when it is impossible under the reaction conditions for the products to be converted back to reagent or converted into one another by any other route:

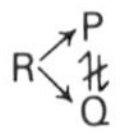

$$R{\nearrow}^{P}_{\searrow Q}$$

(*b*) when it *is* possible under the reaction conditions for interconversion of the products to occur:

$$R \overset{P}{\underset{Q}{\rightleftharpoons}} \quad \text{or} \quad R \overset{P}{\underset{Q}{\rightleftharpoons}}$$

*Case* (*a*). If P and Q cannot be interconverted (via R or otherwise), their relative proportions in the product mixture will depend on the relative rates of the two processes R → P and R → Q. Such a reaction, or rather mixture of reactions, is said to be under *kinetic control* or rate control. The major product will be the one which is formed the fastest. In such cases the activation energies for the two reactions will be important in determining which product is formed.

*Case* (*b*). If P and Q can be interconverted, either by reversion to R or otherwise, so that P and Q are in equilibrium, their relative amounts will depend on the equilibrium constant, that is, on the free energy difference between them, and not on the activation energies for their formation. Reactions of this type are said to be under *thermodynamic control*.

In a reaction under thermodynamic control, the stability (free energy of formation) of the possible products under the prevailing conditions determines which predominates. In reactions under kinetic control, the activation energies of the two reactions determine the ratio of products.

In this book we are largely concerned with the reactions of nucleophiles with electrophiles. Let us examine how the concepts of kinetic and thermodynamic control can be applied to the reactions of a nucleophile.

For a molecule which behaves as a nucleophile (Lewis base), we can define two different characteristics. One, which is a measure of the stability of the bond formed to an electrophile, is called the *basicity* of the nucleophile. It is important in deciding the outcome of reactions which are thermodynamically controlled. The other characteristic of the nucleophile is its *nucleophilicity*, which is a measure of the rate at which it forms a bond to an electrophile. This is important in deciding the outcome of reactions which are kinetically controlled. Ions or molecules can be good nucleophiles (have high nucleophilicity) but be weak bases (have low basicity). The two characteristics are not necessarily connected. The rate of formation of a bond is not directly related to the bond strength. This distinction will be discussed again in the following sections.

### A case of thermodynamic control

The 2-aminoethoxide ion, $^{\ominus}\overline{O}-CH_2-CH_2-\overline{N}H_2$, has two functional groups, both of which have lone pairs of electrons which could be used to form a bond to a proton. If this ion were treated with a source of protons, for example sulphuric acid, it might be converted in to either of two products:

$$^{\ominus}|\overline{O}-CH_2-CH_2-\overline{N}H_2 \quad H^{\oplus} \rightarrow \quad ^{\ominus}|\overline{O}-CH_2-CH_2-\overset{\oplus}{N}H_3$$

or

$$H^{\oplus} \quad ^{\ominus}|\overline{O}-CH_2-CH_2-\overline{N}H_2 \rightarrow \quad H-\overline{O}-CH_2-CH_2-\overline{N}H_2$$

The products differ only in the site of attachment of the proton. They could be interconverted by transfer of the proton back to the solvent and recapture of it at the other end. Most proton transfers are very rapid; for example, the proton transfer involved in the titration of acetic acid with sodium hydroxide is 'instantaneous'. We might expect, therefore, that the protonation of the aminoalkoxide ion should be a reaction in which the products are interconvertible rapidly. In fact, it is. The reaction is therefore under thermodynamic control and the more stable product predominates.

The amino alcohol, $HOCH_2CH_2NH_2$, is by far the major product. We can attribute this to the fact that alkoxides are more basic than amines. As we saw on p. 53, the basicity of an atom depends on the charge and on the electronegativity of the atom. Negatively charged ions are generally stronger bases than neutral molecules with the same donor atom. Amide ($-R_2N^\ominus$), hydroxide, and alkoxide ions are stronger bases than ammonia, amines, water, and alcohols.

You may already have demonstrated that ammonia and amines are liberated from their salts by hydroxide ion. This simply means that hydroxide ion is a stronger base than amines.

$$HO^\ominus + {}^\oplus NH_4 \rightarrow NH_3 + H_2O$$

$$HO^\ominus + R\overset{\oplus}{N}H_3 \rightarrow RNH_2 + H_2O$$

The basicity of the following molecules in fact decreases in the order

$$^\ominus NH_2 > RO^\ominus > HO^\ominus > RNH_2 > NH_3 > H_2O$$

The basicity of an atom is a reflection of the stability of the compound formed when it donates its electrons to form a bond to another atom (see p. 69). Hence the product of protonation of the aminoethoxide ion above on the oxygen atom is more stable than the product of protonation on nitrogen.

The recipient atom in the case above was hydrogen. 'Basicity' usually refers to basicity towards hydrogen, but basicities towards other atoms could be compared in the same way.

### Factors important in thermodynamic control

In a situation where thermodynamic control applies, the outcome depends only on the relative stabilities of the possible product molecules, i.e. on the free-energy difference between the alternative products. The relative stabilities of molecules can be interpreted in terms of—

different bond strengths;

different extents of delocalization;

different arrangements of electrical charges;

different degrees of angle deformation or molecular crowding;

etc., and these factors may be interconnected.

As we have seen, $HOCH_2CH_2NH_2$ is more stable than $^\ominus OCH_2CH_2\overset{\oplus}{N}H_3$ because of the greater strength of an $H-O-$ bond than an $H-N\!\!<$ bond and also because of the energy tied up in a molecule with opposite electrical charges close together.

The ion $^{\oplus}CH_2-\overline{O}-H \leftrightarrow CH_2=\overset{\oplus}{O}-H$ is more stable than its isomer $CH_3-\overline{O}^{\oplus}$ because (among other reasons) the former is delocalized whereas the latter is not. Therefore, the addition under thermodynamic control of a proton to methanal (formaldehyde):

$$CH_2=\overline{O} + H^{\oplus} \rightarrow {}^{\oplus}CH_2-\overline{O}-H \leftrightarrow CH_2=\overset{\oplus}{O}-H$$

yields only the delocalized ion.

### A case of kinetic control

Methyl bromide, $CH_3-Br$, can react with nucleophilic reagents (see p. 83) in a reaction in which the C—Br bond breaks with the formation of a bromide ion and a new bond is formed between the carbon and the nucleophilic atom. For example:

$$H-\overline{O}|^{\ominus} \curvearrowright CH_3-\overline{Br}| \rightarrow H-\overline{O}-CH_3 + |\overline{Br}|^{\ominus}$$

This type of reaction will be discussed in more detail in Chapter 9.

If we treat methyl bromide with the cyclic ether shown below, there are two possible products, both of which arise by this type of reaction. One product, an oxonium ion, would be formed if the oxygen atom were the nucleophilic group and the other product, a sulphonium ion, would be formed if the sulphur atom were the nucleophilic group.

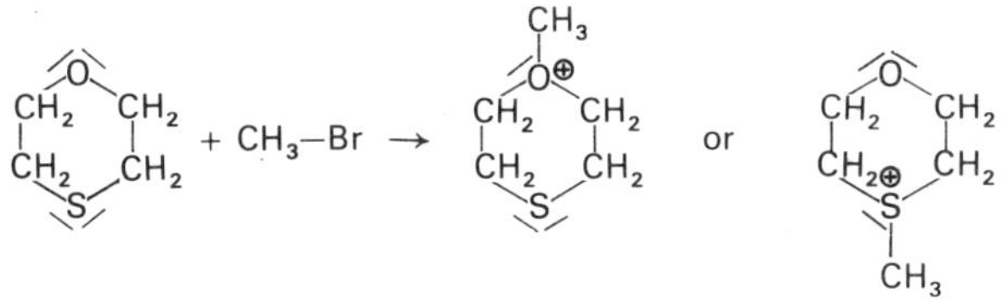

An oxonium ion        A sulphonium ion

*Problem 7–2. Draw mechanisms for the formation of the two products.*

It happens that the products are not interconvertible under the reaction conditions. The reaction is therefore under kinetic control. The observation that the sulphonium ion is the predominant product means that the sulphur atom can donate its electrons to form a bond to the carbon atom of the methyl bromide faster than the oxygen atom can. The activation energy for the reaction of the sulphur is lower than for reaction of the oxygen; or, in other words, the nucleophilicity (see p. 69) of sulphur is greater than that of oxygen.

### Factors important in kinetic control

In a situation where kinetic control applies, the outcome depends on the relative rates of the competing reactions and therefore on the relative energies of the transition states for these reactions. The factors affecting the relative energies of transition states are more numerous and complex than those affecting the relative energies of product molecules and there is as yet no complete theory that allows the prediction of transition-state energies or even relative transition-state energies, although recently progress has been made in predicting which of two pathways from the same reagent pair will have the lower activation energy.

Since the factors determining the relative energies of transition states can be considerably different from those determining the relative energies of products, it should not be surprising that the product that is produced the fastest sometimes is also the most stable product possible, and sometimes it is not.

## NUCLEOPHILES AND ELECTROPHILES AGAIN

Although one of the reactions discussed above is under thermodynamic control and the other under kinetic control, the mechanistic pictures of the two reactions are very similar:

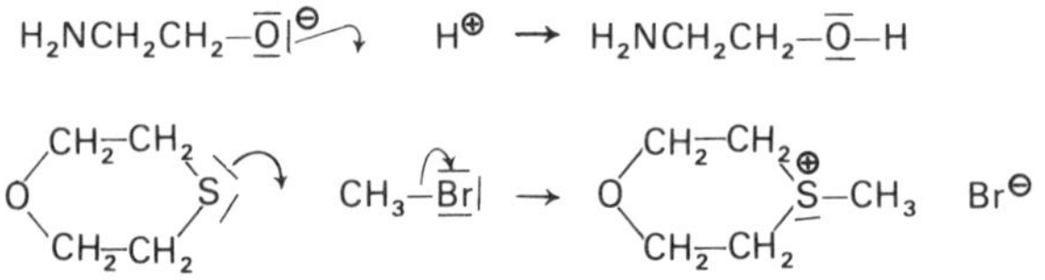

$$H_2NCH_2CH_2-\overline{\underline{O}}|^{\ominus} \quad H^{\oplus} \rightarrow H_2NCH_2CH_2-\overline{\underline{O}}-H$$

If the solvated proton involved in the first case is represented instead by the hydroxonium ion, the similarity is even more marked:

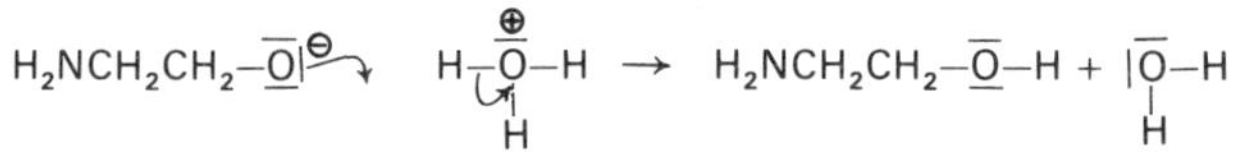

$$H_2NCH_2CH_2-\overline{\underline{O}}|^{\ominus} \quad H-\overset{\oplus}{\underset{\underset{H}{|}}{O}}-H \rightarrow H_2NCH_2CH_2-\overline{\underline{O}}-H + |\overline{\underline{O}}-H$$

In each case one reagent (the nucleophile) donates electrons to form a bond and one reagent (the electrophile) accepts electrons to form a bond. Almost all ionic reactions can be considered as the reaction of a nucleophile with an electrophile. We have here an example of an electrophile ($H^{\oplus}$) which can accept a new bonding pair without any bond breaking and an example of one ($CH_3-Br$) which can accept the new bonding pair only if there is breakage of an existing single bond. A third type is involved in reaction (g) above:

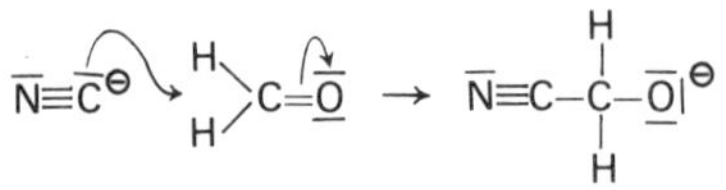

All of the nucleophiles discussed here have donated lone-pair electrons to form the new bond, but ethene can behave as a nucleophile by sharing bonding electrons with another atom (see p. 116):

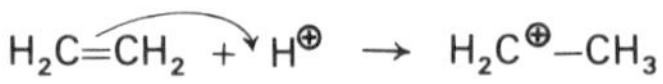

$$H_2C=CH_2 + H^{\oplus} \rightarrow H_2C^{\oplus}-CH_3$$

It is a useful exercise to collect nucleophiles and electrophiles, grouping them according to these types and adding to your list as you read on. It will soon become possible to 'invent' possible reactions that you have not seen by picking out one of each from your lists.

*Problem 7–3. Label the reagents in the following reactions as nucleophiles or electrophiles:*

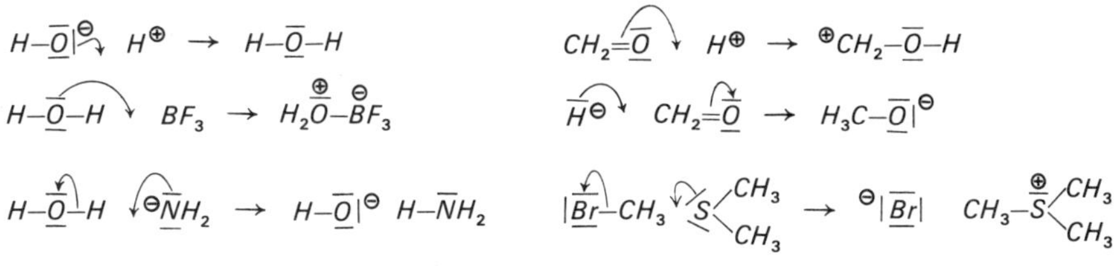

$$H-\overline{\underline{O}}|^{\ominus} \quad H^{\oplus} \rightarrow H-\overline{\underline{O}}-H \qquad CH_2{=}\overline{\underline{O}} \quad H^{\oplus} \rightarrow {}^{\oplus}CH_2-\overline{\underline{O}}-H$$

$$H-\overline{\underline{O}}-H \quad BF_3 \rightarrow H_2\overset{\oplus}{\underline{O}}{-}\overset{\ominus}{\underline{B}}F_3 \qquad H^{\ominus} \quad CH_2{=}\overline{\underline{O}} \rightarrow H_3C-\overline{\underline{O}}|^{\ominus}$$

$$H-\overset{\,}{\underline{O}}{-}H \quad {}^{\ominus}\overline{N}H_2 \rightarrow H-\overline{\underline{O}}|^{\ominus} \quad H-\overline{N}H_2 \qquad |\overline{Br}{-}CH_3 \quad {}^{S}{\overset{CH_3}{\underset{CH_3}{<}}} \rightarrow {}^{\ominus}|\overline{Br}| \quad CH_3{-}\overset{\oplus}{S}{\overset{CH_3}{\underset{CH_3}{<}}}$$

*Problem 7–4. Suggest, in the light of the classification you made above, probable products from the reaction $H^{\ominus} + H{-}O{-}H$. Have you ever carried out this reaction?*

### ANSWERS TO PROBLEMS IN CHAPTER 7

*Problem 7–1.*

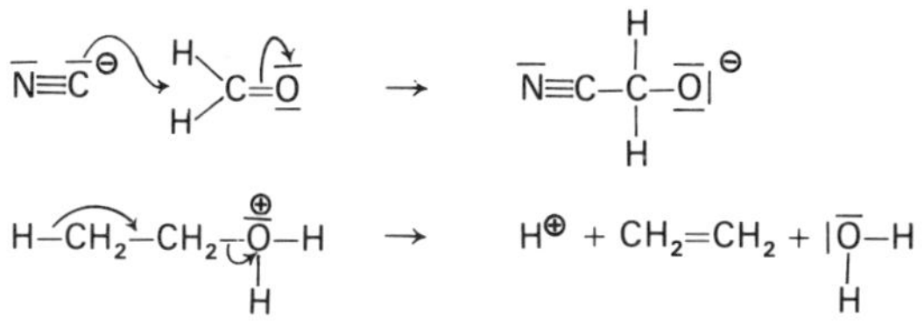

$$\overline{N}{\equiv}C^{\ominus} \quad {\overset{H}{\underset{H}{>}}}C{=}\overline{\underline{O}} \rightarrow \overline{N}{\equiv}C{-}\overset{H}{\underset{H}{C}}{-}\overline{\underline{O}}|^{\ominus}$$

$$H{-}CH_2{-}CH_2{-}\overset{\oplus}{\underset{H}{O}}{-}H \rightarrow H^{\oplus} + CH_2{=}CH_2 + |\overline{\underline{O}}{-}H \atop H$$

*Problem 7–2.*

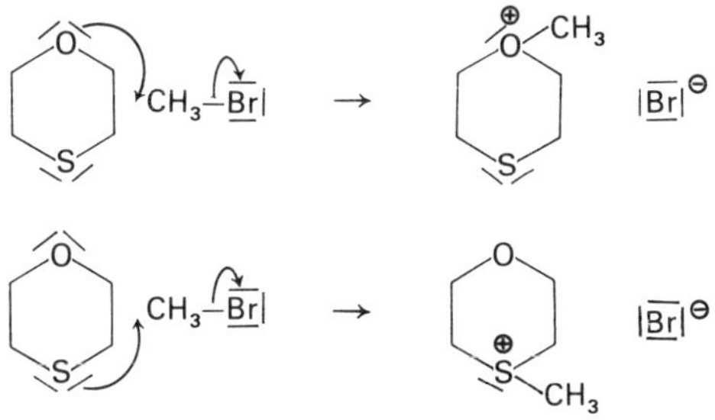

*Problem 7–3.*

$$H-\overline{\underline{O}}|^{\ominus} \qquad H^{\oplus} \rightarrow H-\overline{\underline{O}}-H$$

Nucleophile    Electrophile

$$CH_2{=}\overline{\underline{O}} \quad H^{\oplus} \rightarrow {}^{\oplus}CH_2-\overline{\underline{O}}-H$$

Nucleophile    Electrophile

$$H-\overline{\underline{O}}-H \quad \overset{F}{\underset{F}{\overset{|}{B}}}{-}F \rightarrow \overset{H}{\underset{H}{>}}\overset{\oplus}{\underline{O}}{-}\overset{\ominus}{B}{\overset{F}{\underset{F}{<}}}F$$

Nucleophile    Electrophile

$${}^{\ominus}\overline{H} \quad CH_2{=}\overline{\underline{O}} \rightarrow H-CH_2-\overline{\underline{O}}|^{\ominus}$$

Nucleophile    Electrophile

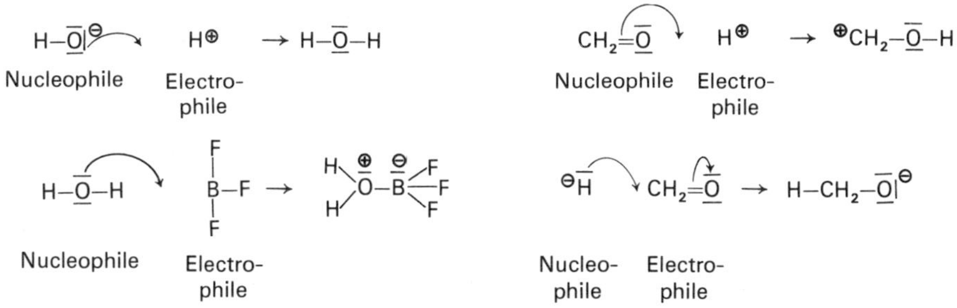

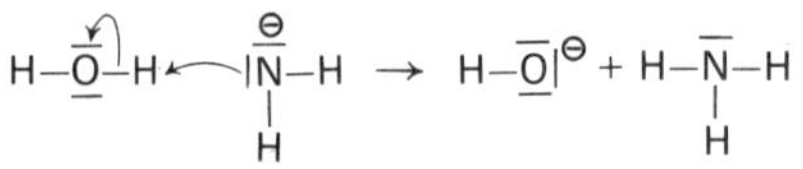

$$H-\overset{}{\underset{|}{\overset{}{O}}}-H \longleftarrow \overset{\ominus}{\overset{}{\underset{|}{N}}}-H \rightarrow H-\overline{O}|^{\ominus} + H-\overset{}{\underset{|}{\overline{N}}}-H \qquad |\overline{Br}-CH_3 \quad \overset{}{S}\overset{CH_3}{\underset{CH_3}{\diagup}} \rightarrow |\overline{Br}|^{\ominus} + CH_3-\overset{\oplus}{\underset{CH_3}{S}}\overset{CH_3}{\diagdown}$$

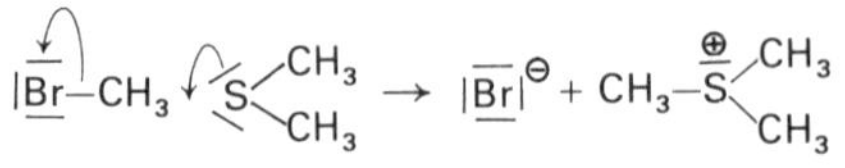

Electrophile  Nucleo-phile                              Electro-phile  Nucleo-phile

Note that molecules such as water or formaldehyde can sometimes behave as nucleophiles and sometimes as electrophiles. You cannot label them until you see how they behave in a given reaction. This topic is taken up again in Chapter 17.

*Problem 7–4.* $\overline{H}^{\ominus}$ can behave only as a nucleophile, donating its lone pair of electrons to form a bond. If any reaction occurs, the water must be behaving as an electrophile. In the examples above, water could act as an electrophile with breakage of an O–H bond. The following reaction is therefore a possibility:

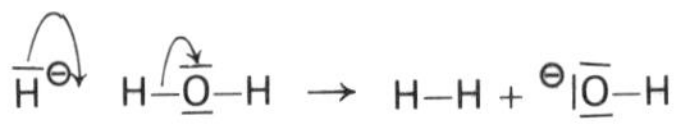

$$\overline{H}^{\ominus} \quad H-\overset{}{\underset{}{O}}-H \rightarrow H-H + {}^{\ominus}|\overline{O}-H$$

This is in fact what happens. The reaction of calcium hydride with water is sometimes used to prepare hydrogen or to dry solvents by destroying trace amounts of water. The proposal

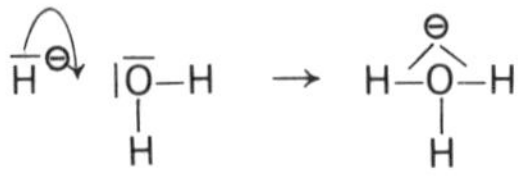

$$\overline{H}^{\ominus} \quad |\overline{O}-H \underset{|}{\underset{H}{}} \rightarrow H-\overset{\ominus}{\underset{|}{\underset{H}{O}}}-H$$

is impossible since oxygen cannot accommodate more than eight valency electrons. The reaction

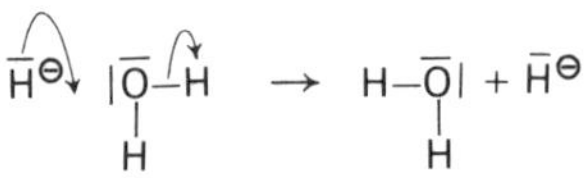

$$\overline{H}^{\ominus} \quad |\overline{O}-H \underset{|}{\underset{H}{}} \rightarrow H-\overline{O}| + \overline{H}^{\ominus} \underset{|}{\underset{H}{}}$$

might happen, but it is very much slower than the rapid proton transfer which is observed.

# Chapter 8

## Alkanes

Molecules constructed only of hydrogen and carbon are called hydrocarbons. Those with no double or triple bonds are said to be *saturated hydrocarbons*, since it is not possible for reagents to add on to them. Those saturated hydrocarbons which are non-cyclic are called alkanes, for example butane and 3-ethyl-2-methylpentane:

$$CH_3-CH_2-CH_2-CH_3$$

$$\begin{array}{c} CH_3 \\ | \\ CH_3-CH_2-CH-CH-CH_3 \\ | \\ CH_3-CH_2 \end{array}$$

Butane                3-Ethyl-2-methylpentane

Alkanes can be branched or unbranched. They all have the generalized formula $C_nH_{2n+2}$.

*Problem 8–1. What is the value of* n *for 3-ethyl-2-methylpentane? Is this compound chiral?*

The naming of these compounds follows the rules given in Chapter 3. The ethyl and methyl groups in the second example are considered as substituent alkyl groups attached to the longest ($C_5$) chain. The drawing above indicates how the correct name of the compound is derived. The molecule could just as well have been drawn as

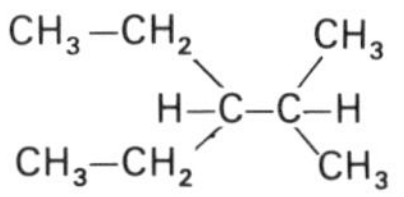

$$\begin{array}{c} CH_3-CH_2 \qquad CH_3 \\ \diagdown \quad \diagup \\ H-C-C-H \\ \diagup \quad \diagdown \\ CH_3-CH_2 \qquad CH_3 \end{array}$$

Make a model of the compound. You will see from the drawing immediately above or from the model that the molecule can adopt a conformation which has a plane of symmetry. The molecule contains four methyl groups ($CH_3$), two methylene groups ($-CH_2-$) and two methine groups ($>CH-$). Looking at it another way, we see two ethyl groups. This fact is not immediately obvious from

75

the correct systematic name. These ethyl groups can be considered as identical groups attached to carbon atom A below and there are two methyl groups which are identical groups attached to carbon atom $B$:

$$\text{two ethyl groups} \left\{ \begin{array}{cc} \vdots\text{CH}_3\text{–CH}_2\vdots & \vdots\text{CH}_3\vdots \\ \text{H–C–C–H} \\ \vdots\text{CH}_3\text{–CH}_2\vdots & \vdots\text{CH}_3\vdots \end{array} \right\} \text{two methyl groups}$$

Cycloalkanes such as cyclopentane and bicycloheptane are also known:

Cyclopentane          Bicycloheptane

Those with one ring all have the formula $C_nH_{2n}$, those with two rings $C_nH_{2n-2}$, and so on. Cycloalkanes have the same types of atoms and bonds as alkanes and show very similar physical and chemical properties.

## BEHAVIOUR AND USES OF SATURATED HYDROCARBONS

Alkanes and cycloalkanes have low boiling points and melting points compared with those of other substances of similar molecular weight.

| Compound | B.p. (°C) | Compound | B.p. (°C) | Compound | B.p. (°C) |
|---|---|---|---|---|---|
| Methane | −162 | Hexane | 69 | $CH_3CH_2CH_3$ | −42 |
| Ethane | −89 | Cyclohexane | 81 | $CH_3$–O–$CH_3$ | −24 |
| Butane | −1 | Decane | 174 | $CH_3CH_2OH$ | 78 |
| 2-Methylpropane | −12 | | | | |

The alkanes of lower molecular weight are gases. Those of higher molecular weight are liquids, oils, greases, or waxes; they are all insoluble in water and they are all rather unreactive at room temperature. They sound rather uninteresting. Can you think of any uses for an inert, water-repelling, greasy liquid?

The decomposition of vegetable matter of past ages has produced vast amounts of alkanes, largely methane and unbranched alkanes of various chain lengths, which have been trapped in spongy rocks below layers or domes of impervious rock such as rock-salt. Drilling in the Middle East, Africa, Venezuela, the U.S.A., the U.S.S.R., Alaska, and the North Sea has made this alkane mixture available to us. We use it in several ways. Small amounts of hydrocarbons are used as lubricants (engine oil, vaseline). This use takes advantage of their inert, greasy and water-repelling character. Some of the alkane mixture is used as a source of hydrogen or carbon to manufacture other substances, such as solvents, plastics, fertilizers, and foodstuffs. About 90% of petroleum and natural gas is burned, which may seem a waste of the most readily available organic material apart from the cellulose and lignin in trees and other plants.

Why are alkanes volatile or greasy, insoluble in water, and inert? The molecules are made of atoms of almost the same electronegativity. The C—C and C—H bonds are therefore non-polar. There are only very weak forces of attraction between any one molecule of the alkane and its neighbour, so the processes of melting and boiling require only small increases in energy. Movement of molecules across or past each other is also easy. There is no attraction between alkane and water molecules comparable to that between water and water molecules, so the water forms a separate phase and stands up in beads on a waxed surface.

The non-polarity of the C—C and C—H bonds also means that these bonds will not readily break to give ions (heterolysis). We saw in Chapter 5 that methane does not dissociate at all and that the methyl anion is a very strong base capable of capturing a proton from almost any other molecule:

$$CH_4 \;\not\rightleftarrows\; {}^{\ominus}\bar{C}H_3 + H^{\oplus}$$

Even if a nucleophile or an electrophile were present to combine with one of the fragments, the other fragment would be a hydride ion, a carbanion, or a carbonium ion—all of which are unstable, reactive ions. Alkanes will therefore react by ionic mechanisms only under forcing conditions. Reactions such as the attack of a nucleophile ($NR_3$) or an electrophile ($BR_3$) on ethane do not occur. The reverse of these proposed reactions occurs readily:

$$\bar{N}R_3 + H-\overset{\displaystyle H}{\underset{\displaystyle H}{C}}-\overset{\displaystyle H}{\underset{\displaystyle H}{C}}-H \;\not\rightleftarrows\; R_3\overset{\oplus}{N}-CH_3 + {}^{\ominus}\bar{C}H_3$$

$$R_3B + H-\overset{\displaystyle H}{\underset{\displaystyle H}{C}}-\overset{\displaystyle H}{\underset{\displaystyle H}{C}}-H \;\not\rightleftarrows\; R-\overset{\displaystyle R}{\underset{\displaystyle R}{B}}{}^{\ominus}H + {}^{\oplus}CH_2CH_3$$

## REACTIONS WITH CHLORINE—CHEMICALS FROM ALKANES

Homolysis of C—C and C—H bonds requires input of almost as much energy as their heterolysis. However, radical reactions (see p. 62), in which a molecule reacts with a radical to produce another molecule and another radical, require less activation energy and are less endothermic, since the total numbers of intact bonds at the beginning and end are now the same:

$$H-\overset{\displaystyle H}{\underset{\displaystyle H}{C}}-H \;\rightarrow\; H-\overset{\displaystyle H}{\underset{\displaystyle H}{C}}\cdot + \cdot H \quad \Delta H \text{ high and positive}$$

whereas

$$X\cdot + CH_4 \;\rightarrow\; \cdot CH_3 + HX \quad \Delta H \text{ relatively small}$$

Reactions of alkanes with radicals are therefore possible at room temperature. An example which is used commercially is the reaction of methane with chlorine. When chlorine is irradiated with light, the molecules can gain energy by absorbing the light and dissociate to chlorine atoms:

$$Cl_2 + light \rightarrow 2Cl\cdot \text{ (strictly } |\overline{Cl}\cdot\text{ )}$$

This process requires less energy than the homolysis of C—C or C—H bonds. These chlorine atoms can react with methane to give HCl and methyl radicals by transfer of a hydrogen atom:

$$Cl\cdot + CH_4 \rightarrow \cdot CH_3 + HCl$$

The methyl radicals could attack any of the molecules or radicals present but the fastest process is attack on a chlorine molecule with chlorine atom transfer:

$$\cdot CH_3 + Cl—Cl \rightarrow CH_3—Cl + Cl\cdot$$

This produces a chlorine atom again, which can react with methane. These two steps can follow one another provided that the supply of reagents lasts and provided that no other reactions occur to trap the reactive radicals. Such a cyclical sequence of steps whose products include a reagent for the next step is called a *chain reaction*:

$$Cl_2 \rightarrow 2Cl\cdot \qquad \text{initial production of a few radicals}$$
$$Cl\cdot + CH_4 \rightarrow HCl + \cdot CH_3 \ \Big\} \ \text{sequence or chain of}$$
$$\cdot CH_3 + Cl_2 \rightarrow CH_3Cl + Cl\cdot \Big/ \ \text{two steps}$$

The sum of the two chain-carrying steps here is the net overall equation

$$Cl_2 + CH_4 \rightarrow CH_3Cl + HCl$$

However, as soon as the chloromethane concentration begins to rise, abstraction of H· from this molecule will compete with abstraction of H· from methane. The radical ·$CH_2Cl$ will be formed and involvement of this in the Cl· transfer step will produce $CH_2Cl_2$. This will not stop the chain. Abstraction of H· from this molecule will also be possible. The result is that chlorination of methane will produce a mixture of chloromethane, dichloromethane, trichloromethane (chloroform), and tetrachloromethane (carbon tetrachloride), all of which are useful substances.

## REACTION WITH OXYGEN—ENERGY FROM ALKANES

Although alkanes are inert to oxygen at room temperature, they do react at higher temperatures and the burning of alkanes provides over one-third of our energy requirements. Natural gas, petrol, and fuel oil keep us warm, move us around, make our electricity and keep our manufacturing industries going. Methane is the main component of the natural gas piped to our homes and factories. Propane and butane are bottled for burning in camps and caravans, boats, and bothies. Petrol, diesel oil, kerosene, paraffin oil, heavy oil and candle wax are all hydrocarbon mixtures which are used as sources of heat.

The combustion of alkanes is another example of a sequence of radical reactions. The radicals required to start the sequence are usually produced by local heating (a match or a spark). The burning of all alkanes is highly exothermic.

$$CH_4 + 2O_2 \rightarrow CO_2 + 2H_2O \quad \Delta H = -882 \text{ kJ}$$

In other words, the heat of combustion of methane is 882 kJ per mole or 55 kJ per gram. The heats of combustion of heptane and of cellulose are 48 and 17 kJ $g^{-1}$, respectively.

*Problem 8–2. Which gives more heat, the burning of an electric fire rated at 1 kW for one hour, the burning of 100 g of petrol or the burning of 200 g of dry cellulose?*

## ALKANES AS SOURCES OF CARBON AND HYDROGEN

It is also possible to oxidize hydrocarbons with oxygen under control to give partly oxidized products. Thus, the exothermic air oxidation of naphtha (a mixture of small hydrocarbons derived from petroleum) is an important industrial process for making acetic acid, $CH_3COOH$, and partial exothermic oxidation of methane by air or its endothermic oxidation by water can be made to yield carbon monoxide and hydrogen:

$$CH_4 + \tfrac{1}{2}O_2 \rightarrow CO + 2H_2$$
$$CH_4 + H_2O \rightarrow CO + 3H_2$$

These are then used to make methanol:

$$CO + 2H_2 \rightarrow CH_3OH$$

and phosgene, which goes to make polyurethane foams:

$$CO + Cl_2 \rightarrow COCl_2$$

and ammonia:

$$3H_2 + N_2 \rightarrow 2NH_3$$

for making fertilizers, such as ammonium sulphate [$(NH_4)_2SO_4$] and urea ($NH_2CONH_2$), and for making nylon:

$$\overset{\displaystyle O}{\underset{\displaystyle \|}{+NH-CH_2CH_2CH_2CH_2CH_2C+}}$$

Nylon-6 polymer

## REACTIONS WITH CATALYSTS—CLOTHES AND FURNITURE FROM ALKANES

The alkanes from petroleum fractions or natural gas can be converted into useful organic molecules by 'cracking'. When the alkanes are heated in contact with suitable catalysts, they are converted into mixtures of small molecules, usually $H_2$, $CH_4$, $CH_2{=}CH_2$, $CH_3CH{=}CH_2$, $HC{\equiv}CH$, and $CH_2{=}CH-CH{=}CH_2$. The alkenes are particularly important since they are used to make plastics [polyethylene, polypropylene, poly(vinyl chloride), etc.] or other bulk chemicals (acetone, acetic acid, etc.). About 20 million tons of ethene are made per year by cracking alkanes.

'Cracking' of petroleum is also carried out in order to convert the larger involatile components into smaller molecules which can be used in petrol. This supplements the small proportion of hydrocarbons of suitable boiling point which is present in the original petroleum.

## REACTION WITH ENZYMES—FOOD FROM ALKANES

Although alkanes are unreactive to most reagents at room temperature, they do react with oxygen at the temperatures in a flame and they do react at about 500 °C on the surface of the catalysts used in 'cracking'. With highly specialized catalysts such as enzymes they can react at room temperature. Certain bacteria and yeasts can grow and multiply even when their only source of carbon is in the form of alkanes. They can oxidize and break down the alkanes and then synthesize fats, sugars, and proteins using only the carbon from this source plus oxygen from the air and nitrogen, sulphur, and phosphorus from inorganic salts. These organisms can therefore convert mineral oil into proteins, which can be used to augment the inadequate world supply of these essential foodstuffs. The detailed chemistry of the enzymic oxidation of alkanes is not yet understood.

## ANSWERS TO PROBLEMS IN CHAPTER 8

*Problem 8–1.* It has the formula $C_8H_{18}$. Therefore $n = 8$. It is not chiral.

| *Problem 8–2.* | The electric fire produces | $1 \times 60 \times 60$ | $= 3600 \text{ kJ}$ |
|---|---|---|---|
| | | | $(1 \text{ W} = 1 \text{ J s}^{-1})$ |
| | The petrol (heptane) produces | $100 \times 48$ | $= 4800 \text{ kJ}$ |
| | The cellulose produces | $200 \times 17$ | $= 3400 \text{ kJ}$ |

# Chapter 9

## Alkyl Halides

Replacement of one hydrogen atom of an alkane by a halogen atom produces a halogenoalkane (alkyl halide):

$CH_3CH_2Cl$

Chloroethane
(ethyl chloride)

Bromocyclopentane
(cyclopentyl bromide)

Replacement of more than one hydrogen by halogen atoms gives molecules which are usually called polyhalogenoalkanes. Most simple halogenoalkanes are liquids and are insoluble in water. Their molecules are of low polarity (see p. 48) and do not form strong hydrogen bonds.

| | Name | | |
| Formula | Alkyl halide | Haloalkane | B.p. (°C) |
|---|---|---|---|
| $CH_3Cl$ | Methyl chloride | Chloromethane | −24 |
| $CH_3I$ | Methyl iodide | Iodomethane | 42 |
| $CH_3CH_2CH_2CH_2Cl$ | — | 1-Chlorobutane | 78 |
| $CH_2Cl_2$ | Methylene chloride | Dichloromethane | 41 |
| $CHCl_3$ | Chloroform | Trichloromethane | 61 |
| $CCl_4$ | Carbon tetrachloride | Tetrachloromethane | 77 |

*Problem 9–1. How many structural and stereoisomers with the molecular formula $C_4H_9F$ are there? What are their systematic names?*

Alkyl halides are generally reactive substances that are useful in syntheses, as we shall see later. Polyhalogenoalkanes are generally unreactive and find

81

applications which take advantage of this property. For example, $CF_2Cl_2$ (b.p. $-30\ ^\circ$C) is used as the circulating fluid in refrigerators and as the propellant in aerosol cans. What would happen if the can were punctured or put in an incinerator?

Carbon tetrachloride is used as a 'dry' cleaning fluid and as a fire extinguisher. The dense vapour of the $CCl_4$ forms a layer above the fire, excluding oxygen from the combustible material. However, at the temperatures reached in a fire, or even in a lit cigarette, the carbon tetrachloride reacts slightly with oxygen to give the corrosive and toxic gas phosgene, $COCl_2$. Smoking while dry-cleaning your clothes and use of carbon tetrachloride to extinguish fires in enclosed spaces are therefore dangerous practices.

Poly(tetrafluoroethylene), PTFE or 'Teflon', is an inert, waxy polymer which is hydrophobic (water-repellant) like polyethylene. It is used as a coating for skis and frying pans and for making bearings and seals and gaskets.

$$+CF_2CF_2CF_2CF_2CF_2CF_2CF_2+_n$$

Poly(tetrafluoroethylene)

We are concerned in this chapter primarily with the reactions of alkyl halides. Did you find four compounds of formula $C_4H_9F$? One of them is chiral.

| $CH_3CH_2CH_2CH_2F$ | $(CH_3)_2CHCH_2F$ | $\overset{\displaystyle CH_3}{\underset{\displaystyle}{CH_3CH_2CHF}}$ | $(CH_3)_3CF$ |
|---|---|---|---|
| 1-Fluorobutane, | 1-Fluoro-2-methylpropane | 2-Fluorobutane, | 2-Fluoro-2-methylpropane |
| a primary alkyl halide | a primary alkyl halide | a secondary alkyl halide | a tertiary alkyl halide |

Each of them contains a fluoro group and an alkyl group. The alkyl groups can be classified according to the nature of the substituents on the first carbon atom, that is, the one carrying the halogen atom. Primary alkyl groups have two hydrogen and one carbon atom attached to the first carbon atom, $C-\overset{\displaystyle H}{\underset{\displaystyle H}{C}}-$. Methyl groups are also primary alkyl groups. Secondary alkyl groups have two carbon and one hydrogen atom, $C-\overset{\displaystyle H}{\underset{\displaystyle C}{C}}-$. Tertiary alkyl groups have three carbon atoms attached to the first carbon atom, $C-\overset{\displaystyle C}{\underset{\displaystyle C}{C}}-$.

1-Fluorobutane and 1-fluoro-2-methylpropane are therefore primary alkyl halides. 2-Fluorobutane is a secondary alkyl halide and 2-fluoro-2-methylpropane is a tertiary one. This classification is sometimes used to provide names for compounds. Thus, 2-fluorobutane is sometimes called secondary butyl fluoride and 2-fluoro-2-methylpropane is sometimes called tertiary butyl fluoride.

*Problem 9–2. Label the alkyl groups in the following compounds as primary, secondary or tertiary alkyl groups:*

$$CH_3CH_2-\overset{\overset{\displaystyle CH_3}{|}}{\underset{\underset{\displaystyle Br}{|}}{C}}-CH_3$$

$$CH_3CH_2-O-CH(CH_3)_2$$

Certain halogenoalkanes can be made by free-radical halogenation of alkanes, as discussed in the last chapter:

$$CH_4 + Br_2 \rightarrow CH_3Br + HBr$$

Other preparative routes start from alcohols or alkenes and will be discussed in Chapters 11 and 13.

Alkyl halides differ from alkanes in possessing lone pairs of electrons on the halogen and a carbon–halogen bond. Because of the greater electronegativity of most halogens compared with that of carbon (see p. 48), alkyl halides such as methyl bromide are slightly polar substances while methyl iodide is almost non-polar. In spite of this, methyl iodide is a more reactive substance than its analogues methyl bromide and methyl chloride. The reason for this may lie in the relative strengths of carbon–halogen bonds. It requires 330 kJ mol$^{-1}$ to break $CH_3-Cl$ into $CH_3\cdot$ and $Cl\cdot$ but only 220 kJ mol$^{-1}$ to break $CH_3-I$ into $CH_3\cdot$ and $I\cdot$. Although the reactions discussed below involve breaking the bonds to produce halide ions rather than radicals, it is reasonable to suppose that breaking a C–I bond is always easier than breaking a C–Cl bond.

## REACTIONS OF THE CARBON–HALOGEN BOND

### Substitution reactions

Alkyl halides can react with nucleophiles in a substitution reaction in which the halogen is liberated as halide ion and the nucleophile becomes bonded to the carbon atom of the alkyl group. Such substitutions at tetrahedral carbon atoms by nucleophiles with the displacement of a leaving group, which takes the bonding pair with it, are called $S_N$ *reactions*. Ethyl bromide reacts with iodide ions in a suitable solvent to give ethyl iodide and bromide ions:

$$CH_3CH_2-Br + I^{\ominus} \rightarrow CH_3CH_2-I + Br^{\ominus}$$

*Problem 9–3. What properties should a suitable solvent for this reaction have?*

The electrons of the C–Br bond end up as a lone pair on the bromide ion and the iodide ion contributes one of its lone pairs to form a bond to the carbon. The mechanism could be depicted as

We are not concerned here with the question of whether the two electron pair shifts are simultaneous. We shall assume that they are, although this is not the case for all $S_N$ reactions. Nor are we concerned with the shape of the transition state.

This $S_N$ reaction seems to be a possible pathway for the replacement of halogen atoms by any nucleophile and indeed we shall see later that halogens are not the only groups which can be displaced in this way. Many different kinds of nucleophiles might be used. The mechanism is the same in each case:

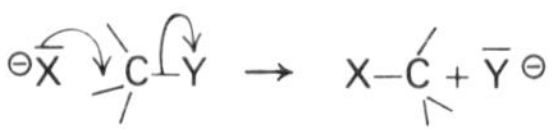

$$^{\ominus}X \quad C \quad Y \quad \rightarrow \quad X-C + \bar{Y}^{\ominus}$$

Hydroxide ions can react with alkyl halides to give alcohols:

$$(CH_3)_2CHCH_2Br + HO^{\ominus} \rightarrow (CH_3)_2CHCH_2OH + Br^{\ominus}$$

Alkoxide ions give ethers:

$$CH_3Br + CH_3CH_2O^{\ominus} \rightarrow CH_3-O-CH_2CH_3 + Br^{\ominus}$$

Ethoxide ion

Carboxylate ions give esters:

$$CH_3CH_2I + CH_3-C{\overset{\displaystyle O}{\underset{\displaystyle O^{\ominus}}{}}} \rightarrow CH_3CH_2-O-\overset{\displaystyle O}{\overset{\displaystyle \|}{C}}-CH_3 + I^{\ominus}$$

Acetate ion

Hydrosulphide ions give thiols, the sulphur-containing analogues of alcohols:

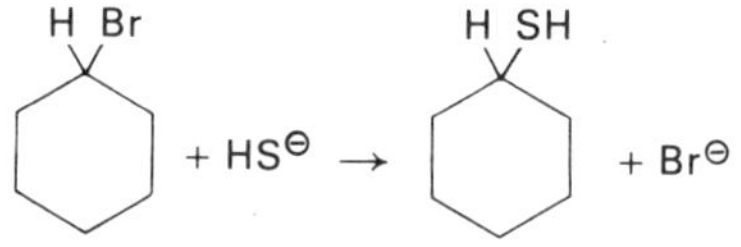

Bromocyclohexane    Cyclohexanethiol

The syllable -thio- appears in several chemical names and indicates that an oxygen atom has been replaced by sulphur. For example:

$$NCO^{\ominus} \qquad NCS^{\ominus} \qquad SO_4^{2-} \qquad S_2O_3^{2-}$$

Cyanate ion   Thiocyanate ion   Sulphate ion   Thiosulphate ion

Carbanions form new C—C bonds:

$$CH_3CH_2I + {}^{\ominus}CN \rightarrow CH_3CH_2CN + I^{\ominus}$$

Ethyl cyanide
(propanonitrile)

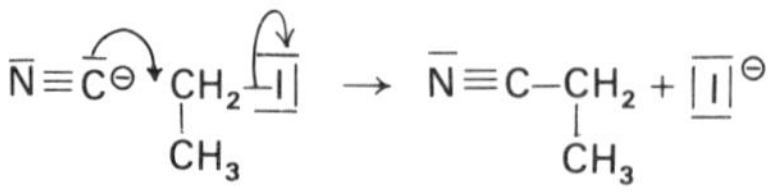

$$\bar{N} \equiv C^{\ominus} \, CH_2 \, I \rightarrow \bar{N} \equiv C-CH_2 + I^{\ominus}$$
$$\qquad\qquad | \qquad\qquad\qquad |$$
$$\qquad\qquad CH_3 \qquad\qquad CH_3$$

and amines form ammonium salts:

$$CH_3I + NH_3 \rightarrow CH_3-\overset{\oplus}{N}H_3 + I^{\ominus}$$

Methylammonium
iodide

Subsequent reaction of the ammonium ion with a base would form an amine by proton transfer:

$$CH_3-\overset{\oplus}{N}H_3 + {}^{\ominus}OH \rightarrow CH_3-NH_2 + H_2O$$
Methylamine

All of these conversions can be made to proceed. Although all of these reactions are in theory reversible, the equilibrium constants are large or the reaction can be pushed to the right by increasing the reagent concentration.

The $S_N$ reactions are usually performed in polar organic solvents which can dissolve both the ionic and the covalent reagents. The solvent must not itself be an effective nucleophile. The reaction of ethyl bromide with lithium iodide could be carried out in acetone and that between methyl bromide and ethoxide ion in ethanol.

This might seem surprising since these solvents have lone pairs of electrons and are potential nucleophiles. Why does the alkyl bromide not react with the alcohol? In fact it can, but much more slowly than with the iodide ion, ammonia, or ethoxide ion. Water and alcohols are much slower acting nucleophiles than amines, thiols, and iodide ion. The neutral oxygen atom in the water loses electrons as the reaction proceeds and eventually develops a full positive charge. Since oxygen is fairly electronegative, this loss of electrons and development of positive charge in the transition state will be more difficult than the analogous reaction of the negatively charged, and less electronegative, iodide ion.

$$H_2\ddot{O} \quad \overset{\curvearrowright}{}C-\overset{\curvearrowright}{Br}| \rightarrow H_2\overset{\oplus}{O}-C + |\overline{Br}|^{\ominus}$$

$$\overset{\ominus}{\underline{I}} \quad \overset{\curvearrowright}{}C-\overset{\curvearrowright}{Br}| \rightarrow \underline{I}-C + |\overline{Br}|^{\ominus}$$

We can put nucleophiles in a rough order of *nucleophilicity* (rate of reaction)—rough because the order can be altered if we change the solvent or if we change the electrophile. Anions, $A^{\ominus}$, are better nucleophiles than the related neutral molecule, HA. Halide ions, alkoxide, hydroxide, and carbanions are good nucleophiles, as are ammonia, amines, and thiols. Water, alcohols, ethers, phenols, carboxylic acids, alkenes, and benzene are poorer nucleophiles.

The reaction of neutral nucleophiles of the type HA such as alcohols and water with alkyl halides will liberate protons which will be solvated:

$$R-Br + H_2O \rightarrow R-\overset{\oplus}{O}H_2 + Br^{\ominus} \rightarrow ROH + H^{\oplus} + Br^{\ominus}$$

and, as we shall see later, this reaction, like the ones above, is reversible. If the protons are captured by a base which is not itself a good nucleophile, and which can be put in the reaction mixture without itself reacting with the alkyl halide, the equilibrium can be displaced to give the product.

The reaction of phenol (hydroxybenzene) with ethyl bromide in acetone containing (insoluble) potassium carbonate is an interesting and preparatively useful example:

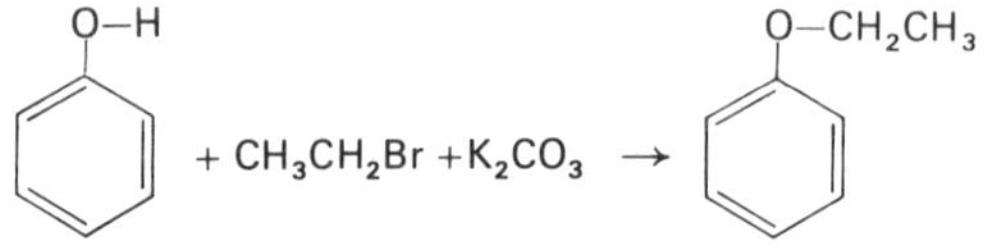

Phenol

*Problem 9–4. Complete the stoichiometric equation for the above reaction.*

$S_N$ reactions of alkyl halides can be used to convert them into many other types of compounds. The reaction involves loss of a halide ion with attack at carbon by a nucleophile. Many nucleophiles can be used successfully.

*Problem 9–5. What would be the products of $S_N$ reactions between the following reagents?:*

(a) $CH_3CH_2CH_2Br + KCN$

(b) $CH_3-CH-CH_3 + LiI$
     |
     $Cl$

(c) $+ H_2NCH_3$

(d) $HOCH_2CH_2NH_2 + CH_3-I$

## Elimination reactions

In the reactions of alkyl halides we have discussed so far, the halogen atom behaves as a leaving group. It is lost from the organic molecule as a halide ion and the nucleophile becomes bonded to the carbon in its place.

Most alkyl halides can also react with nucleophiles in a different way—again with displacement of halide ion but with attack by the nucleophile occurring at hydrogen instead of at carbon:

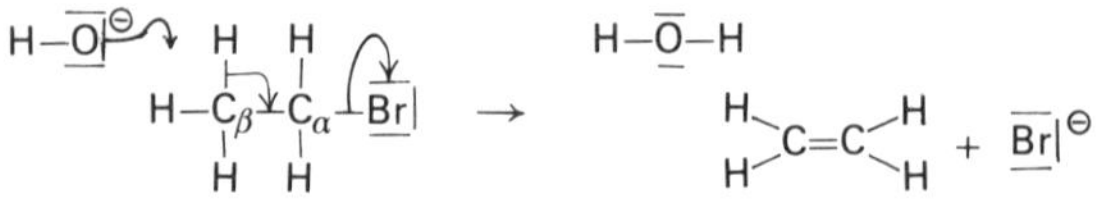

This new process is called an elimination or $E$ reaction. The bromide ion and one hydrogen ion are eliminated or lost from the alkyl halide and an alkene is formed. The proton is usually captured by the reagent nucleophile. The proton removed has to be one attached to the carbon atom (called the $\beta$-carbon) next to the one carrying the halogen. Removal of any other proton may be possible but would not lead to an alkene. If there are no hydrogen atoms on any of the $\beta$-carbon atoms then elimination to give an alkene is not possible. In ethyl bromide there are three

$\beta$-hydrogen atoms. In 1-bromo-1-methylcyclohexane there are seven, and two different alkenes could be formed.

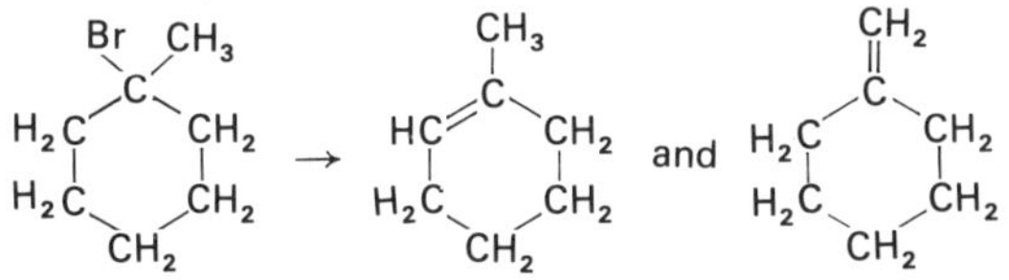

An alkyl halide such as ethyl bromide can react with a nucleophile such as hydroxide in at least two different ways—by an $S_N$ process to give ethanol and bromide ion or by an $E$ process to give ethene, bromide ion, and water. When the reagents are mixed, either or both paths could be followed. Neither reaction is reversible under the usual conditions, so the predominant product is that from the faster reaction (kinetic control). The ratio of substitution products to elimination products depends on the relative rates at which the nucleophile used attacks the hydrogen or the carbon atom of the halide, and that depends on the natures of all of the reagents. The important fact is that two pathways are possible.

*Problem 9–6. What products would you expect from the following reactions?:*

$$(CH_3)_3C–CH_2Br + NaOH$$

$$(CH_3)_2CH–\underset{\underset{Br}{|}}{C}HCH_3 + NaOH$$

### Reduction

There is one other important reaction of alkyl halides which again results in conversion of the halogen into halide ion. Alkyl halides are reducible by various reagents, including electropositive metals such as lithium. The products are lithium ion, halide ion, and a covalent alkyllithium:

$$CH_3–I + 2Li \quad \rightarrow \quad Li^{\oplus} + I^{\oplus} + CH_3–Li$$

Methyllithium

Check that the total number of electrons is the same before and after. The lithium metal is oxidized to $Li^{\oplus}$ and to monocovalent lithium. The $CH_3I$ is reduced.

A similar reaction occurs with magnesium:

$$CH_3–I + Mg \quad \rightarrow \quad CH_3–Mg–I$$

The product contains a covalent C—Mg bond. This reaction was discovered by Victor Grignard (pronounced 'Greenyarr') and the alkylmagnesium halides, such as methylmagnesium iodide, are often called Grignard reagents. Some reactions of organometallic compounds such as methyllithium and methylmagnesium bromide are discussed in Chapters 10 and 15. Other reductions of alkyl halides, by radical pathways, also occur but are not discussed here.

## REACTIONS INVOLVING THE LONE-PAIR ELECTRONS

All of the reactions of alkyl halides described so far have involved the carbon–halogen bond in the first step. However, reactions are known in which the lone-pair electrons on the halogen atom are involved first. Because of the presence of these electrons, alkyl halides might behave as nucleophiles or Lewis bases. They do in fact interact with electrophiles such as aluminium chloride forming complexes:

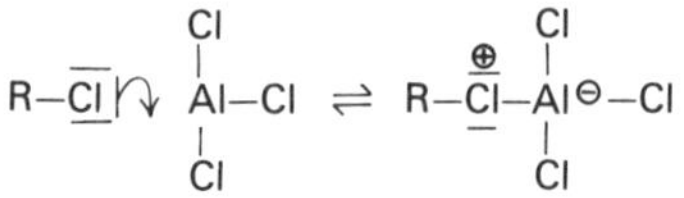

$$R-\overline{\underline{Cl}} \curvearrowright \overset{Cl}{\underset{Cl}{Al}}-Cl \rightleftharpoons R-\overset{\oplus}{\underset{\underline{Cl}}{Cl}}-\overset{Cl}{\underset{Cl}{Al}}{}^{\ominus}-Cl$$

which can be represented by the dipolar structure shown. The R–Cl bond in this complex is now very easily broken to give either a carbonium ion or a product resulting from simultaneous attack by a nucleophile:

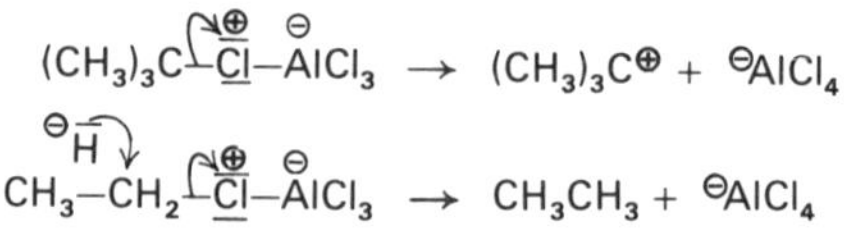

$$(CH_3)_3C \overset{\curvearrowleft \oplus}{\underset{}{\underline{Cl}}}-AlCl_3 \rightarrow (CH_3)_3C^{\oplus} + {}^{\ominus}AlCl_4$$

$$\overset{\ominus \overset{}{H}}{\underset{CH_3-CH_2}{\curvearrowright}} \overset{\curvearrowleft \oplus}{\underline{Cl}}-\overset{\ominus}{AlCl_3} \rightarrow CH_3CH_3 + {}^{\ominus}AlCl_4$$

It is therefore possible to bring about $S_N$ reactions of alkyl halides with certain nucleophiles, which are insufficiently reactive by themselves, by treating the halide first with an electrophile such as $AlCl_3$ or $Ag^{\oplus}$. Some further examples will be discussed in Chapter 14.

## ANSWERS TO PROBLEMS IN CHAPTER 9

*Problem 9–1.* See p. 82.

*Problem 9–2.*

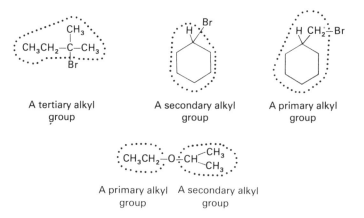

| A tertiary alkyl group | A secondary alkyl group | A primary alkyl group |
|---|---|---|

A primary alkyl group   A secondary alkyl group

*Problem 9–3.* The solvent should be moderately polar in order to dissolve both reagents and should not itself be a good nucleophile. See p. 85.

*Problem 9–4.*

$$C_6H_5OH + C_2H_5Br + K_2CO_3 \rightarrow C_6H_5OC_2H_5 + KBr + KHCO_3$$

*Problem 9–5.*

(a) $CH_3CH_2CH_2CN + KBr$

(b) $CH_3-CH-CH_3 + LiCl$
          |
          I

(c)

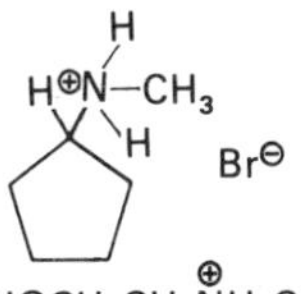

(d) $HOCH_2CH_2\overset{\oplus}{N}H_2CH_3 \ I^{\ominus}$

Of the two nucleophilic groups, the nitrogen reacts faster than the oxygen. See pp. 71 and 85.

*Problem 9–6.*

$(CH_3)_3C-CH_2OH$. Substitution only because there are no $\beta$-hydrogen atoms, so elimination is not possible.

$(CH_3)_2C=CHCH_3 \ + \ (CH_3)_2CH-CH=CH_2 \ + \ (CH_3)_2CH-\underset{\underset{OH}{|}}{C}HCH_3$

Elimination, involving either of two different $\beta$-hydrogen atoms, and substitution are all possible.

# Chapter 10

# Organometallic Compounds

There is a wide range of compounds which contain carbon to metal bonds. In this chapter, only those in which the carbon–metal bond is a simple single covalent bond are discussed, and the word metal is taken to include all elements less electronegative than carbon, such as boron and silicon as well as lithium, copper, thallium, lead, etc.

## METHODS OF MAKING COMPOUNDS WITH CARBON–METAL BONDS

There are three major methods of forming carbon–metal bonds: (a) from the metal and an alkyl halide, (b) by substitution of one metal for another or for a proton, and (c) by addition of a compound of the metal to an alkene.

(a) Alkyl metals such as methyllithium and ethylmagnesium bromide can be made by reduction of the alkyl halide by the metal, as we saw in the last chapter:

$$CH_3-I + 2Li \rightarrow CH_3-Li + Li^\oplus + I^\ominus$$

(b) Treatment of these alkylmetals with different metal ions may result in metal exchange. For example, methyllithium can react with Cu(I) salts to give methylcopper:

$$CH_3-Li + Cu^\oplus \rightarrow CH_3-Cu + Li^\oplus$$

and Grignard reagents can react with silicon halides to form new C–Si bonds:

$$CH_3-Mg-Br + SiCl_4 \rightarrow CH_3-SiCl_3 + MgBrCl$$

It may also be possible to exchange a proton for a metal ion, but since most C–H bonds do not dissociate readily, it is necessary to add a base in order to remove the proton.

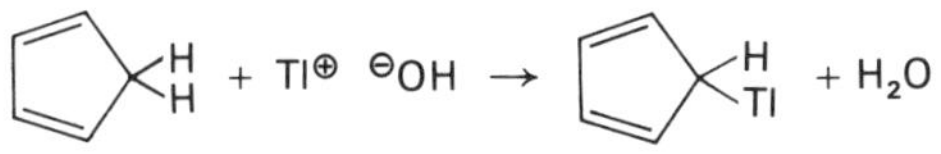

Cyclopentadiene   Thallium(I) hydroxide   Cyclopentadienylthallium

$$H{-}C{\equiv}C{-}H + CH_3{-}Mg{-}Br \rightarrow H{-}C{\equiv}C{-}Mg{-}Br + CH_4$$

Ethyne (acetylene)   Ethynylmagnesium bromide

In the latter example the methyl Grignard reagent provides the base and the new metal atom.

(*c*) The third important method of making carbon–metal bonds is by additions to alkenes of some simple compound containing the metal:

$$CH_2{=}CH_2 + BH_3 \rightarrow \left[H{-}CH_2CH_2{-}B{\genfrac{}{}{0pt}{}{H}{H}}\right] \xrightarrow{2C_2H_4} CH_3CH_2{-}B{\genfrac{}{}{0pt}{}{CH_2CH_3}{CH_2CH_3}}$$

Borane (prepared as the dimer, $B_2H_6$)   Not isolatable   Triethylborane

The hydride of boron is easily made. It exists as the dimer, diborane ($B_2H_6$), but its reactions can be drawn as if it were simply $BH_3$. It reacts with alkenes such as ethene by adding across the double bond. The process is called hydroboration of the alkene. Addition reactions to alkenes are common and are discussed in Chapter 13, although the mechanisms of the reactions discussed there are not the same as the mechanism of the borane addition. The first product still has two B–H bonds and so can react with two more ethene molecules to give eventually triethylborane.

## PROPERTIES OF ORGANOMETALLIC COMPOUNDS

By variations of these routes, it is possible to prepare alkylmetals from almost any metal. The properties of the alkylmetals vary considerably. Some, such as tetramethylsilane, $(CH_3)_4Si$, are stable. Some, such as trimethylaluminium, $(CH_3)_3Al$, are very easily oxidized or are even spontaneously flammable in air. Some are useful commercially, such as tetraethyllead, which is a valuable but toxic additive put into petrol to modify the burning behaviour of the hydrocarbons.

In synthetic organic chemistry, the heterolytic reactions of the alkylmetals are useful. In all of these compounds a carbon atom is attached to an atom which is more electropositive than it is. The bonds will all be polarized in the sense

$$\overset{\delta-}{\geq C}{-}\overset{\delta+}{M}$$

and heterolysis of the bond will tend to give rise to a metal cation and a carbanion,

92

which may be stable in solution in a suitable solvent or may be simultaneously captured by a reagent electrophile:

$$CH_3\!-\!Li \;\longrightarrow\; {}^{\ominus}CH_3 + Li^{\oplus}$$

Even if the carbanion can exist as such for some time, it will be able to react with electrophiles. Reactions with electrophiles form the major part of the chemistry of alkylmetals.

The electrophile could be a metal ion:

$$Cu^{\oplus} + CH_3\!-\!Li \;\longrightarrow\; Cu\!-\!CH_3 + Li^{\oplus}$$

This is the process discussed above as a method of making alkylcoppers.

Alternatively, the electrophile could be a proton or a proton donor, X—H:

$$X\!-\!H + CH_3\!-\!Li \;\longrightarrow\; \overline{X}^{\ominus} + CH_4 + Li^{\oplus}$$

For example, Grignard reagents react vigorously with water to give hydrocarbons:

$$H\!-\!\underline{O}\!-\!H + CH_3\!-\!Mg\!-\!Br \;\longrightarrow\; H\underline{O}|^{\ominus} + CH_4 + {}^{\oplus}MgBr$$

The process mentioned above for making ethynyl Grignard reagents is similar. Ethyne is the proton donor:

$$H\!-\!C \equiv C\!-\!H + CH_3\!-\!Mg\!-\!Br \;\dashrightarrow\; H\!-\!C \equiv C\!-\!MgBr + CH_4$$

Ethyne            Ethynylmagnesium bromide

Also, the electrophile could be an electrophilic carbon atom:

$$|\overline{Br}\!-\!CR_3 + CH_3\!-\!Cu \;\longrightarrow\; |\overline{Br}|^{\ominus} + CR_3\!-\!CH_3 + Cu^{\oplus}$$

$$\underline{O}\!=\!CR_2 + CH_3\!-\!Mg\!-\!Br \;\longrightarrow\; {}^{\ominus}|\overline{O}\!-\!\overset{R}{\underset{R}{C}}\!-\!CH_3 + {}^{\oplus}MgBr$$

These reactions are all substitutions at carbon by electrophiles or $S_E$ reactions. The metal ion is the leaving group and leaves the bonding pair behind. Only one electron pair is involved in the substitution. Comparison with the $S_N$ reactions of alkyl halides is valuable.

Alkylmetals are nucleophilic. They undergo substitution at carbon when treated with electrophiles such as $H^{\oplus}$.

Alkyl halides are electrophilic. They undergo substitution at carbon when treated with nucleophiles such as $Br^{\ominus}$.

$$S_E: Li\!-\!CH_3\;\;H^{\oplus} \;\longrightarrow\; Li^{\oplus} + H_3C\!-\!H \qquad \text{1 electron pair is involved}$$

$$S_N: |\overline{Br}^{\ominus}\;CH_3\!-\!|\underline{I}| \;\longrightarrow\; |\overline{Br}\!-\!CH_3 + |\underline{I}|^{\ominus} \qquad \text{2 electron pairs are involved}$$

Nucleophile     Electrophile

These mechanistic pictures are not meant to imply anything about the shape of the transition states. The actual mechanisms of both $S_E$ and $S_N$ reactions are sometimes simple, as these pictures suggest, and sometimes more complicated.

Several of the preparations of alkylmetals also involve $S_E$ reactions and further examples will be found in Chapter 15.

Like alkyl halides, alkylmetals are valuable reagents in syntheses, as the metal can be exchanged for many other elements such as metals, hydrogen, or carbon as in the examples given above. The fact that alkyl halides can be converted by reduction into alkylmetals adds enormously to the synthetic potential of the halides.

*Problem 10–1. Construct a diagram as follows and add to it as many interconversions as you can.*

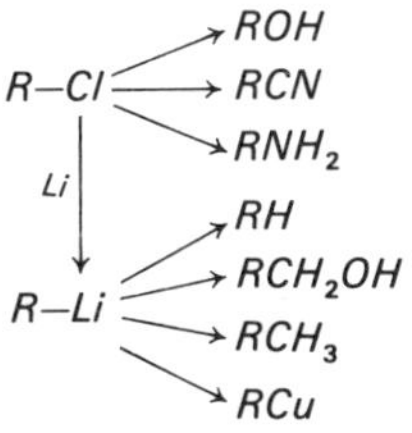

$$
\begin{array}{c}
R\text{–}Cl \longrightarrow
\begin{array}{l}
ROH \\
RCN \\
RNH_2
\end{array} \\[1em]
\Big\downarrow Li \\[1em]
R\text{–}Li \longrightarrow
\begin{array}{l}
RH \\
RCH_2OH \\
RCH_3 \\
RCu
\end{array}
\end{array}
$$

*Problem 10–2. Predict the products from:*

(a) $CH_3CH_2MgBr + CO_2$

(b) $CH_3CH_2Li + H_2O$

(c) $(CH_3CH_2)_3B + H\text{–}O\text{–}\overset{\overset{\displaystyle O}{\|}}{C}\text{–}CH_3$ in water

(d) (cyclopentadienyl) $\overset{H}{\underset{Tl}{\diagup\hspace{-0.3em}\diagdown}} + CH_3I$

(e) $B_2H_6 +$ (cyclododecatriene) $(C_{12}H_{18})$

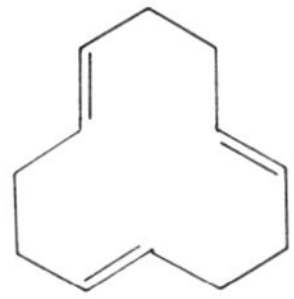

## ANSWERS TO PROBLEMS IN CHAPTER 10

*Problem 10–2.*

(a) $CH_3\text{–}CH_2\text{–}MgBr \;\longrightarrow\; CH_3\text{–}CH_2\text{–}C\overset{\displaystyle \overline{O}}{\underset{\displaystyle \underline{O}|^{\ominus}}{\diagup}} \;\;^{\oplus}MgBr$

(with curved arrows from $CH_3$–$CH_2$ and $O{=}C{=}O$)

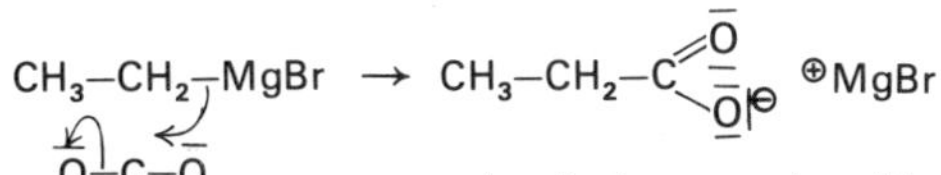

A salt of propanoic acid

The reaction is an $S_E$ reaction at the $CH_2$ of the Grignard reagent. The carbon of the $CO_2$ is electrophilic.

94

(b) $CH_3-CH_2-Li \rightarrow CH_3CH_3 + Li^\oplus + {}^\ominus\overline{O}H$

$H-\overline{O}-H$  Ethane

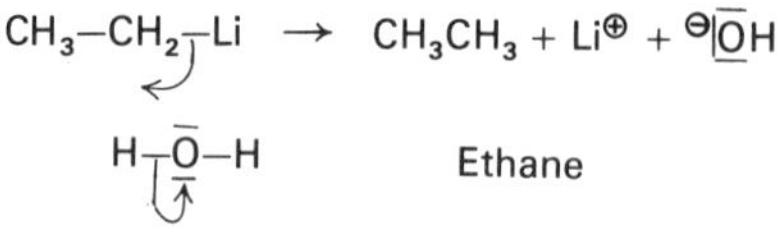

(c) The acetic acid is a good proton donor. Each C—B bond becomes C—H and the boron ends up as $B(OH)_3$.

$$(CH_3CH_2)_3B + H_2O \xrightarrow[\text{catalyst}]{HO\overset{\displaystyle O}{\overset{\|}{C}}CH_3} 3CH_3CH_3 + B(OH)_3$$

(d) + $Tl^\oplus$ + $\overline{\overline{I}}{}^\ominus$

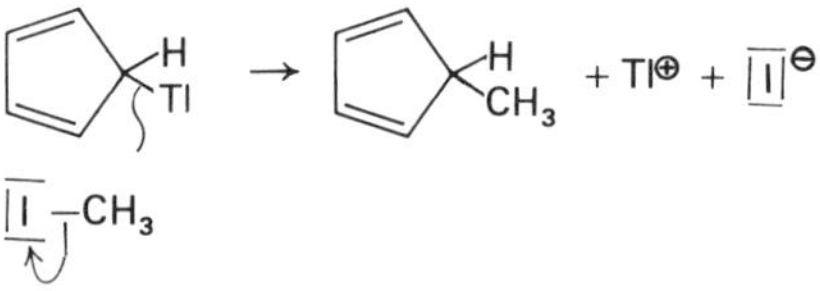

(e) The reaction is like that of borane with ethene but the three double bonds are all in one molecule. The ends of the double bonds are equivalent so only one initial adduct is possible.

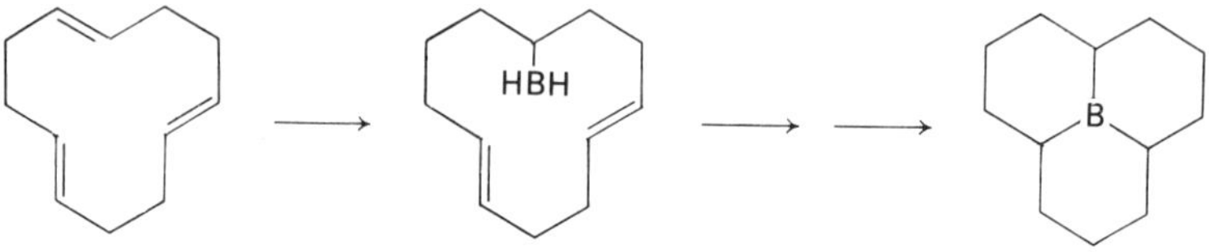

$C_{12}H_{21}B$

# Chapter 11

## Alcohols and Ethers

Organic compounds in which carbon is singly bonded to oxygen are of two types, alcohols, R—O—H, and ethers, which may be symmetrical, R—O—R, or unsymmetrical, R—O—R'. Because of the relatively high electronegativity of oxygen, alcohols are fairly polar molecules which can be attracted to one another and to other polar substances by dipole–dipole attractions and also by hydrogen bonding.

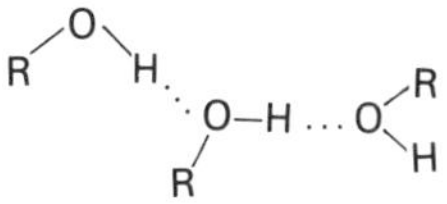

Alcohols of low molecular weight, such as methanol, ethanol, and propan-2-ol, are soluble in water and are liquids which are much less volatile than propane. Alcohols with larger alkyl groups such as cyclohexanol are not very soluble in water since the hydrophobic (water-repelling) hydrocarbon group forms a larger part of the whole molecule. Polyhydroxy compounds such as ethane-1,2-diol (ethylene glycol) and glucose, on the other hand, are very soluble.

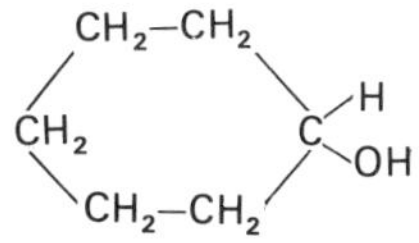

Cyclohexanol

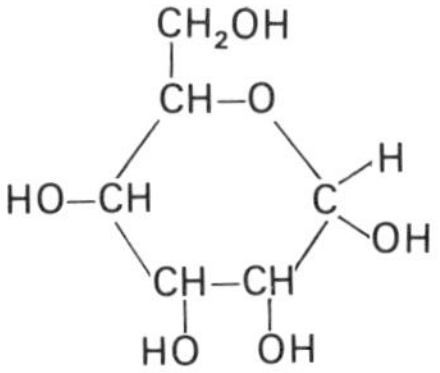

Glucose

Methanol and ethanol are used extensively as solvents, for the preparation of esters (Chapter 16), and to a small extent as fuels. Methanol is used as the carbon feedstock in one process for the growing of bacterial protein (cf. p. 80). Ethanol is a widely consumed drug which depresses mental functions. Large doses can cause addiction or death. Methanol is highly toxic if drunk by animals or people

Physical properties of some representative alcohols

| Formula | Systematic name | Other name | B.p. (°C) | Solubility in water |
|---------|-----------------|------------|-----------|---------------------|
| $CH_3OH$ | Methanol | Methyl alcohol | 65 | Completely soluble |
| $CH_3CH_2OH$ | Ethanol | Ethyl alcohol | 78 | Completely soluble |
| $CH_3CH_2CH_2OH$ | Propan-1-ol | n-Propanol | 97 | Soluble |
| $CH_3CHCH_3$<br>$\quad\mid$<br>$\quad OH$ | Propan-2-ol | Isopropanol | 82 | Soluble |
| $C_6H_{11}OH$ | Cyclohexanol | | 162 | Slightly soluble |
| $C_6H_5CH_2OH$ | Phenylmethanol | Benzyl alcohol | 205 | Slightly soluble |
| $HOCH_2CH_2OH$ | 1,2-Dihydroxy-ethane | Ethylene glycol | 197 | Completely soluble |
| $HOCH_2CHCH_2OH$<br>$\quad\quad\mid$<br>$\quad\quad OH$ | 1,2,3-Trihydroxy-propane | Glycerol | 290 | Completely soluble |

since it is oxidized to formaldehyde (methanal), which combines with bio-chemically important substances. Methanol is oxidized commercially to formal-dehyde for making synthetic resins.

Ethers are moderately polar but can form hydrogen bonds only to molecules with OH, NH, or SH bonds:

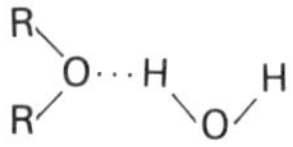

They are generally more volatile and rather less water-soluble than the alcohols of which they are isomers.

| | |
|---|---|
| $CH_3CH_2OCH_2CH_3$ | B.p. 35 °C; solubility 7 g in 100 ml of water. |
| $CH_3CH_2CH_2CH_2OH$ | B.p. 117 °C; solubility 9 g in 100 ml of water. |

Diethyl ether is a useful solvent for organic molecules since it is moderately polar, easily evaporated (b.p. 35 °C) and only sparingly soluble in water. It is interesting that the cyclic ether called tetrahydrofuran is also a good solvent, which is freely soluble in water perhaps because the oxygen is more accessible.

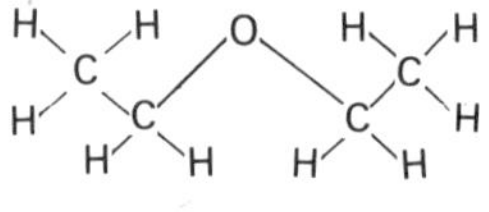

Diethyl ether, $C_4H_{10}O$ 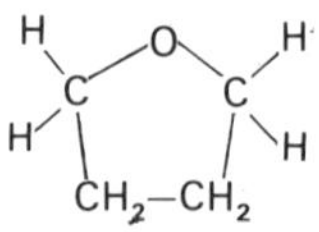          Tetrahydrofuran, $C_4H_8O$

Before discussing the reactions of alcohols, it would be useful to point out the features present in their structure and try to make some predictions. (1) The oxygen atom carries two lone pairs of electrons, so alcohols might behave as nucleophiles. (2) There is a polar C—O bond which is likely, given suitable circumstances, to break in the manner shown:

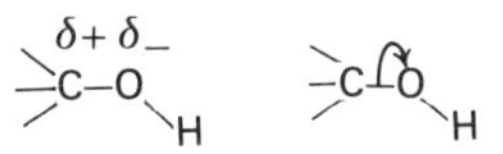

$$\overset{\delta+\ \delta-}{\underset{H}{\overset{}{\searrow}}} \quad \searrow C-O \diagdown H \qquad \searrow C \curvearrowright O \diagdown H$$

(3) There is a polar O—H bond which is likely, given suitable circumstances, to break as shown:

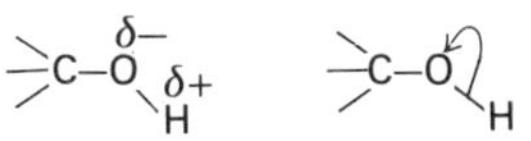

$$\searrow C-\overset{\delta-}{O} \underset{H}{\overset{}{\searrow}} \overset{}{\underset{\delta+}{}} \qquad \searrow C-O \curvearrowright H$$

All of these potential reactions have been observed.

## ALCOHOLS AS NUCLEOPHILES—REACTIONS OF THE LONE-PAIR ELECTRONS

Alcohols react with protons to give oxonium ions:

$$R-\overset{\displaystyle \bar O |}{\underset{H}{\phantom{O}}} + H^\oplus \rightarrow R-\overset{\oplus}{\underset{\underset{H}{|}}{O}}-H$$

in the same way as water does. They are about as basic as water is. We shall discuss the reactions of oxonium ions later in this chapter.

*Problem 11–1. Before we do, can you predict the kinds of reactions that*

$$CH_3CH_2-\overset{\oplus}{\underset{\underset{H}{|}}{O}}-H \text{ might undergo?}$$

Alcohols are not good enough nucleophiles to displace halide ion from alkyl halides (see p. 85), but they do react with better carbon electrophiles such as $R_2C=\overset{\oplus}{O}H$ and $R-C\overset{\diagup O}{\diagdown Cl,}$ as we shall see in Chapters 15 and 16.

*Problem 11–2. Predict the product of reaction of $CH_3OH$ with $(CH_3)_2C=\overset{\oplus}{O}H$*

## ALCOHOLS AS ELECTROPHILES—REACTIONS OF THE C–O BOND

Heterolysis of the C—O bond of alcohols would give a carbonium ion and a hydroxide ion, but this does not occur even in polar solvents. Heterolysis of the bond with simultaneous attack by a nucleophile does not occur either. Indeed, the opposite process occurs readily, as we saw on p. 84:

$$|\overline{Cl}|^\ominus \ CH_3-\overline{O}-H \rightleftarrows |\overline{Cl}|-CH_3 + {}^\ominus|\overline{O}|-H$$

We must conclude that the OH group makes a poor leaving group or, more

98

explicitly, that the oxygen does not have a sufficiently strong pull on the bonding electrons to allow an $S_N$ displacement of hydroxide ion.

It can be made to pull harder, for instance, by giving it a full positive charge. This can be done by converting the alcohol into an oxonium ion by adding acid. The positively charged oxygen is easily displaced.

$$\overline{|Br|}^{\ominus} \curvearrowright CH_3 \overset{\oplus}{\overset{|}{O}} -H \;\rightarrow\; \overline{|Br}-CH_3 + H-\underline{\overline{O}}-H$$
$$\qquad\qquad\quad \underset{H}{|}$$

Treatment of alcohols with acids such as HBr, which provide both the proton and the nucleophile, leads to the alkyl halide:

$$(CH_3)_3C-OH + HCl \;\rightarrow\; (CH_3)_3C-Cl + H_2O$$

The mechanism involves protonation and then $S_N$ displacement:

$$(CH_3)_3C-\underline{\overline{O}}-H \xrightarrow{H^{\oplus}} (CH_3)_3C-\overset{\oplus}{\underset{H}{\overset{|}{O}}}-H \xrightarrow{Cl^{\ominus}} (CH_3)_3C-\overline{\underline{Cl}|} + H-\overline{O}-H$$

Not every nucleophile, however, can be used to replace the hydroxyl group of an alcohol by this protonation–substitution sequence. Alkyloxonium ions may be attacked at the carbon atom by nucleophiles such as $Cl^{\ominus}$, $I^{\ominus}$, or $HSO_4^{\ominus}$, which are not basic:

$$CH_3OH \rightarrow CH_3-\overset{\oplus}{\underset{H}{\overset{|}{O}}}-H \xrightarrow{Br^{\ominus}} CH_3-Br + H_2O \qquad \text{Substitution}$$

However, if the nucleophile is even weakly basic, it will capture a proton from the acidic oxonium ion and reform the original alcohol. Nucleophiles such as ammonia, amines, alkoxide ions, cyanide ion, or carbanions, all of which are basic, cannot be used to replace the OH group of alcohols by other groups:

$$CH_3-OH \rightarrow CH_3-\overset{\oplus}{\underset{H}{\overset{|}{O}}}-H \xrightarrow{{}^{\ominus}CN} CH_3OH + HCN \qquad \begin{array}{l}\text{Proton}\\ \text{transfer}\end{array}$$

Since the $-{}^{\oplus}OH_2$ group of alkyloxonium ions is a good leaving group which can be displaced in an $S_N$ reaction, we must consider whether elimination reactions are also possible. If there is a hydrogen atom on the $\beta$-carbon atom, an elimination reaction will be possible as a competitor to substitution:

$$\underset{\overset{|}{\underset{OH_2^{\oplus}}{}}}{\overset{H}{>}C{\curvearrowright}C<}$$

The situation is analogous to that with alkyl halides. In that case we were able to favour elimination by using a strong base to remove the $\beta$-hydrogen atom. We cannot do the same with oxonium ions since the base would remove the most acidic proton which is one on oxygen:

$$B + \overset{H}{\underset{CH_2CH_2-\overset{|\oplus}{\underset{H}{O}}-H}{|}} \rightarrow {}^{\oplus}BH + \overset{H}{\underset{CH_2CH_2-\underline{\overline{O}}-H}{|}}$$

We can still discourage substitution by using an anion such as the hydrogen sulphate ion, $HSO_4^{\ominus}$, which will not readily form a bond to carbon. If the mixture is heated, until something happens, elimination may be the only possibility. Indeed, warming alcohols with strong acids which have non-nucleophilic anions (for example, sulphuric acid) leads to elimination to give alkenes. The protons act as a catalyst, being involved in the reaction and then regenerated at the end.

$$CH_3CH_2OH + H^{\oplus} \longrightarrow H_2C\overset{\underset{|}{H}}{-}CH \longrightarrow H_2C=CH_2 + \overline{O}H_2 + H^{\oplus}$$
$$\overset{|}{O}H_2^{\oplus}$$

The hydrogen sulphate ion, however, can form a bond to carbon very slowly and some sulphate ester may be formed as a by-product:

$$CH_3-CH_2\overset{\ominus}{-}O-\overset{O}{\underset{O}{\overset{\|}{S}}}-OH \longrightarrow CH_3-CH_2-\overline{O}-\overset{O}{\underset{O}{\overset{\|}{S}}}-OH + H_2\overline{O}$$
$$\overset{|}{O}H_2^{\oplus}$$

Ethyl hydrogen sulphate<br>(an ester)

The elimination of water from an alcohol to form an alkene is called *dehydration* of the alcohol. Acid catalysis is one way of effecting dehydration.

## ALCOHOLS AS PROTON DONORS—REACTIONS OF THE O–H BOND

Alcohols, like water, can donate protons to strong bases:

$$H-O-H + {}^{\ominus}CH_3 \rightleftharpoons H-O^{\ominus} + CH_4$$
Hydroxide<br>ion

$$R-O-H + {}^{\ominus}CH_3 \rightleftharpoons R-O^{\ominus} + CH_4$$
An alkoxide<br>ion

These reactions, like most proton transfers, are reversible and an equilibrium will be set up. The equilibrium constant will be large, that is, formation of the alkoxide ion will be fairly complete, only if the base used is much stronger than the alkoxide ion. Alcohols do not transfer protons to water. Aqueous solutions of alcohols are neutral. They will transfer protons to hydroxide ion to give an equilibrium mixture:

$$ROH + {}^{\ominus}OH \rightleftharpoons RO^{\ominus} + HOH$$

and they are virtually completely deprotonated by strong bases such as $NH_2^{\ominus}$ and $CH_3^{\ominus}$:

$$ROH + {}^{\ominus}NH_2 \longrightarrow RO^{\ominus} + NH_3$$

Alkoxide ions can also be made by reduction of alcohols by metals such as sodium. The reaction is analogous to the reduction of water.

$$2H_2O + 2Na \rightarrow 2HO^{\ominus} + H_2 + 2Na^{\oplus}$$
$$2ROH + 2Na \rightarrow 2RO^{\ominus} + H_2 + 2Na^{\oplus}$$

Sodium ethoxide, $Na^{\oplus}{}^{\ominus}OC_2H_5$, is a useful base which can be made in this way for use under anhydrous conditions. Alkoxide ions are slightly stronger bases than hydroxide ions and they can also readily form bonds to carbon (see p. 84).

## OXIDATION OF ALCOHOLS

Alcohols which contain the group $\diagdown C \diagup ^{H}_{OH}$ can be oxidized to carbonyl compounds containing the group $\diagdown C = O$. The net reaction is a dehydrogenation, i.e. loss of two hydrogen atoms. Dehydrogenation is one form of oxidation. In the case below the hydrogen atoms end up as protons.

Many strong oxidizing agents can oxidize alcohols. Mild oxidizing agents such as $O_2$ or $Cu^{2+}$ have no effect at room temperature. One of the most useful oxidizing agents for alcohols is chromium(VI) oxide, $CrO_3$. The oxidizing power of this reagent can be modified by changing the solvent and the acidity of the medium. The mechanism of the oxidation involves addition of the alcohol to the $CrO_3$ to form a chromate ester, a compound which still contains Cr(VI), which then decomposes to give the carbonyl compound. The chromium acquires the bonding electrons and so is reduced to a compound of Cr(IV):

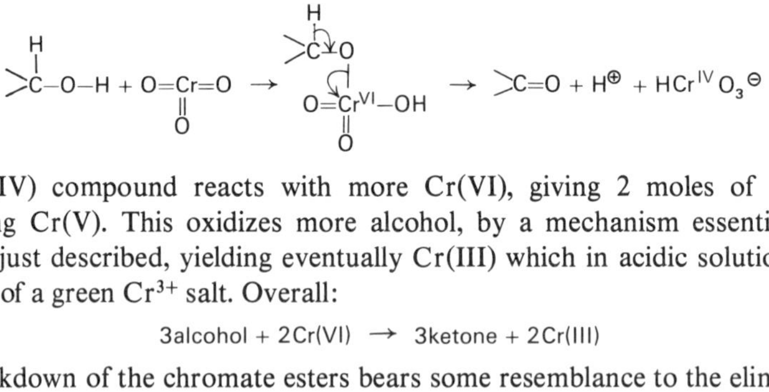

The Cr(IV) compound reacts with more Cr(VI), giving 2 moles of an acid containing Cr(V). This oxidizes more alcohol, by a mechanism essentially the same as just described, yielding eventually Cr(III) which in acidic solution takes the form of a green $Cr^{3+}$ salt. Overall:

$$3\text{alcohol} + 2Cr(VI) \rightarrow 3\text{ketone} + 2Cr(III)$$

The breakdown of the chromate esters bears some resemblance to the elimination of HBr from alkyl halides to form a carbon–carbon double bond:

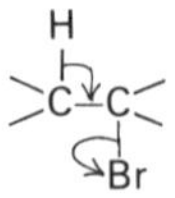

You will see that oxidation of primary alcohols, i.e. those in which the hydroxyl group is attached to a primary alkyl group, will give aldehydes on oxidation. These aldehydes may be further oxidized to acids. Secondary alcohols on oxidation will give ketones. Tertiary alcohols have no hydrogen atom on the carbon atom carrying the hydroxyl group. They may be oxidized under vigorous conditions but only as a result of breakage of a C–C bond.

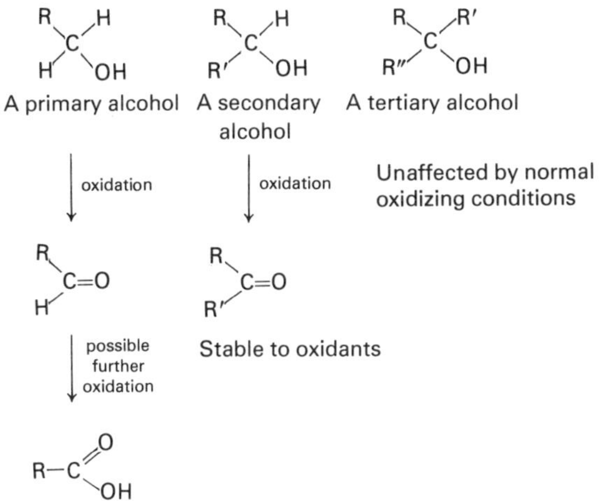

A primary alcohol    A secondary    A tertiary alcohol
                     alcohol

| oxidation | oxidation | Unaffected by normal oxidizing conditions |

possible further oxidation    Stable to oxidants

## ETHERS

Ethers contain the group R—O—R'. Like alcohols, they can be protonated by strong acids and they are very feeble nucleophiles (at oxygen) and very feeble electrophiles (at carbon). They cannot be oxidized or reduced easily, and are rather inert. The dialkyloxonium ions formed by protonation of ethers behave like the alkyloxonium ions discussed above (p. 98). They can be attacked by non-basic nucleophiles such as iodide ion with cleavage of a C—O bond.

$$CH_3-O-C_6H_5 \xrightarrow{HI} CH_3I + HOC_6H_5$$

Methyl phenyl ether                Phenol

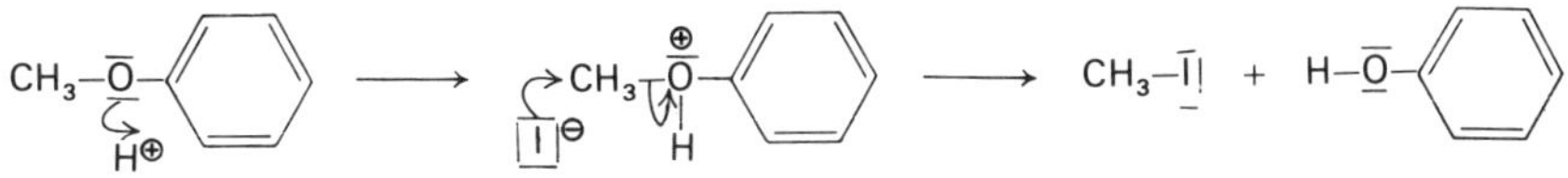

Note that the nucleophilic iodide ion attacks the tetrahedral carbon atom of the methyl group and not the trigonal carbon atom of the phenyl group. No methanol or iodobenzene is formed.

*Problem 11–3. What products would you obtain from the following reactions?:*

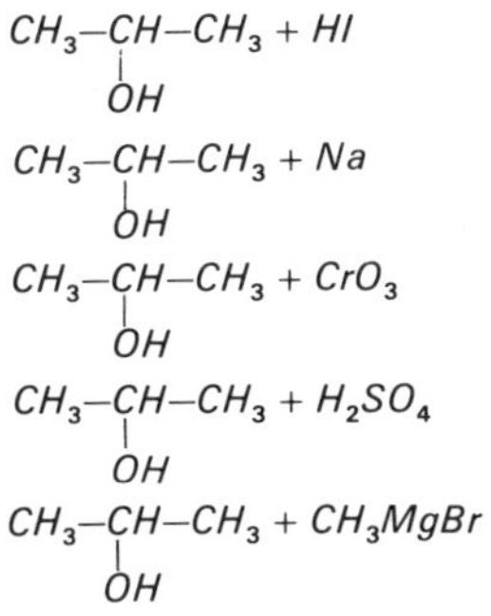

$$CH_3-CH-CH_3 + HI$$
$$| $$
$$OH$$

$$CH_3-CH-CH_3 + Na$$
$$| $$
$$OH$$

$$CH_3-CH-CH_3 + CrO_3$$
$$| $$
$$OH$$

$$CH_3-CH-CH_3 + H_2SO_4$$
$$| $$
$$OH$$

$$CH_3-CH-CH_3 + CH_3MgBr$$
$$| $$
$$OH$$

*Problem 11–4. How could you make* $CH_3-O-CH(CH_3)_2$ *starting from methanol and propan-2-ol?*

## ANSWERS TO PROBLEMS IN CHAPTER 11

*Problem 11–1.* The positive oxygen will strongly attract the electrons of the OH and OC bonds. The oxonium ion should be a good proton donor and should also be able to undergo substitution reactions and elimination reactions.

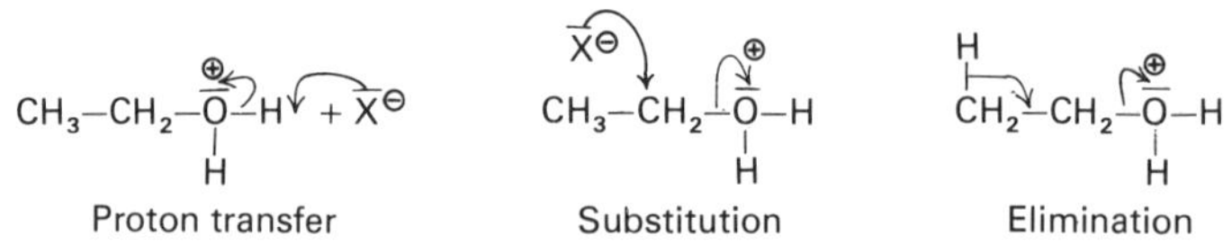

$$CH_3-CH_2-\overset{\oplus}{O}\!-\!H + X^{\ominus} \qquad CH_3-CH_2-\overset{\oplus}{O}-H \qquad CH_2-CH_2-\overset{\oplus}{O}-H$$

Proton transfer $\qquad\qquad$ Substitution $\qquad\qquad$ Elimination

*Problem 11–2.*

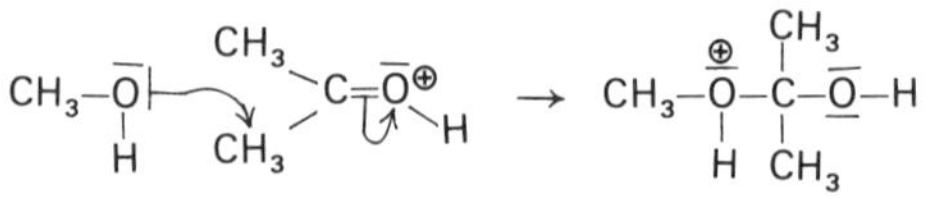

$$CH_3-\overset{-}{O}\!-\!H \quad \overset{CH_3}{\underset{CH_3}{>}}C\!=\!\overset{\oplus}{O}\!-\!H \quad \rightarrow \quad CH_3-\overset{\oplus}{O}-\overset{CH_3}{\underset{CH_3}{\overset{|}{C}}}-\overset{}{O}\!=\!H$$

The protonated ketone can accept electrons only at the carbon atom.

*Problem 11–3.*

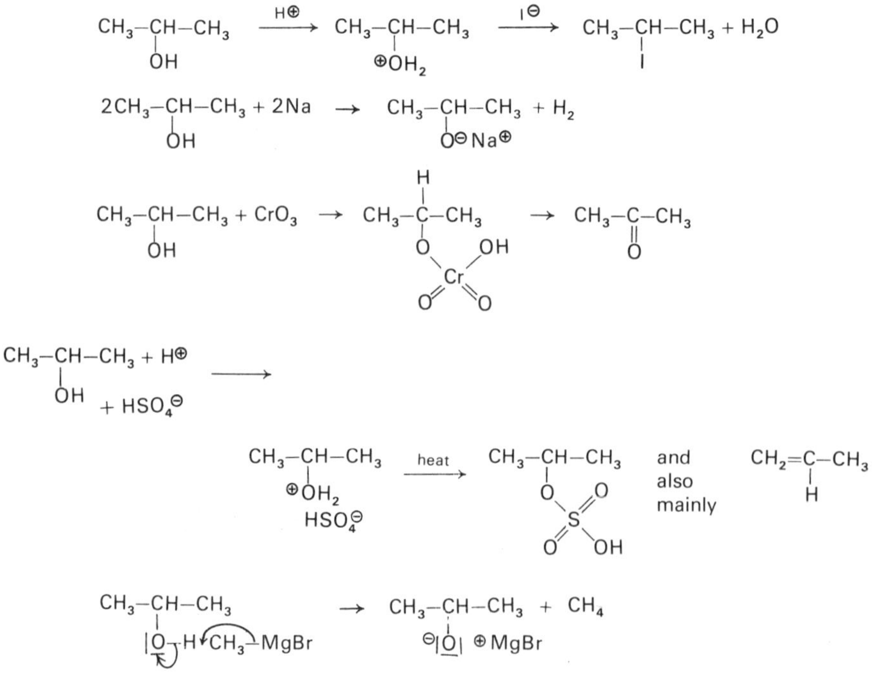

$$CH_3-\underset{OH}{\overset{|}{C}H}-CH_3 \xrightarrow{H^{\oplus}} CH_3-\underset{\oplus OH_2}{\overset{|}{C}H}-CH_3 \xrightarrow{I^{\ominus}} CH_3-\underset{I}{\overset{|}{C}H}-CH_3 + H_2O$$

$$2CH_3-\underset{OH}{\overset{|}{C}H}-CH_3 + 2Na \rightarrow CH_3-\underset{O^{\ominus}Na^{\oplus}}{\overset{|}{C}H}-CH_3 + H_2$$

$$CH_3-\underset{OH}{\overset{|}{C}H}-CH_3 + CrO_3 \rightarrow CH_3-\overset{H}{\underset{O\diagdown{}{Cr}{}\diagup OH}{\overset{|}{C}}}-CH_3 \rightarrow CH_3-\overset{}{\underset{O}{\overset{\|}{C}}}-CH_3$$

$$CH_3-\underset{OH}{\overset{|}{C}H}-CH_3 + H^{\oplus}$$
$$+ HSO_4^{\ominus} \longrightarrow$$

$$CH_3-\underset{\oplus OH_2}{\underset{HSO_4^{\ominus}}{\overset{|}{C}H}}-CH_3 \xrightarrow{heat} CH_3-\overset{}{\underset{O\diagdown{}{S}{}\diagup O}{\overset{|}{C}H}}-CH_3 \quad \begin{matrix}\text{and}\\ \text{also}\\ \text{mainly}\end{matrix} \quad CH_2=\underset{H}{\overset{|}{C}}-CH_3$$

$$CH_3-\overset{}{\underset{O-H\,CH_3\,MgBr}{\overset{|}{C}H}}-CH_3 \rightarrow CH_3-\underset{\ominus O^{\ominus}\oplus MgBr}{\overset{|}{C}H}-CH_3 + CH_4$$

*Problem 11–4.* We will need to make one alcohol into the alkoxide ion and the other into the halide. We then want to bring about an $S_N$ reaction and avoid elimination. Elimination would not be possible if we mixed $CH_3I$ and $^\ominus O–CH(CH_3)_2$. Therefore:

$$CH_3OH + HI \rightarrow CH_3I + H_2O$$

Isolate and purify the $CH_3I$.

$$(CH_3)_2CHOH + Na \rightarrow (CH_3)_2CH–O^\ominus \ Na^\oplus + H_2$$

The reaction is easily carried out in excess of alcohol as solvent. The hydrogen will bubble out. Treat the solution of the alkoxide ion with the methyl iodide:

$$(CH_3)_2CH–O^\ominus \ Na^\oplus + CH_3–I \rightarrow (CH_3)_2CH–O–CH_3 + Na^\oplus \ I^\ominus$$

Distil out the ether to separate it from the excess of propan-2-ol or pour the reaction mixture into water once all of the $CH_3I$ has reacted. All of the substances will dissolve except the ether, which will form a separate layer.

# Chapter 12

# Amines

Amines are related to ammonia much as alcohols and ethers are related to water. The naming of them is a little complex. They are grouped into three types according to the number of alkyl groups present:

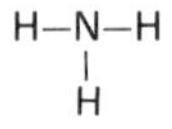

$$\underset{\text{Ammonia}}{\overset{\displaystyle H-N-H}{\underset{\displaystyle |}{\overset{\displaystyle |}{H}}}} \qquad \underset{\text{A primary amine}}{\overset{\displaystyle R-N-H}{\underset{\displaystyle |}{\overset{\displaystyle |}{H}}}} \qquad \underset{\substack{\text{A secondary}\\\text{amine}}}{\overset{\displaystyle R-N-R'}{\underset{\displaystyle |}{\overset{\displaystyle |}{H}}}} \qquad \underset{\text{A tertiary amine}}{\overset{\displaystyle R-N-R'}{\underset{\displaystyle |}{\overset{\displaystyle |}{R''}}}}$$

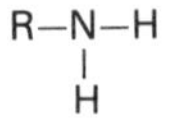

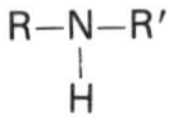

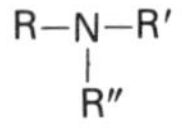

The nature of the alkyl groups does not affect this classification, only their number.

*Problem 12–1. Can you imagine a compound with four alkyl groups attached to nitrogen? Which simple ion will it be related to?*

The alkyl groups in the amines may, of course, be primary, secondary or tertiary alkyl groups (see p. 82). Thus, 2-aminopropane is a primary amine containing a secondary alkyl group and piperidine is a secondary amine with two primary alkyl groups attached to the nitrogen atom. Hence the words primary, etc., are used in two ways: (*a*) to describe the number of alkyl groups on the nitrogen and (*b*) to describe these alkyl groups.

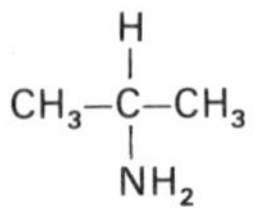

| 2-Aminopropane | Piperidine | Aniline (phenylamine) |

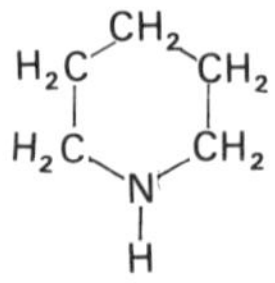

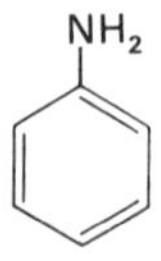

Some amines have special names such as piperidine and aniline. Aniline is the common name for aminobenzene (phenylamine). It is a primary amine. There is yet another way of naming amines. Aminoethane can be called ethylamine and this system is used for other amines such as diethylamine. $N$-Methylethylamine, in which a methyl group replaces one of the hydrogen atoms on the $N$itrogen atom of ethylamine is a secondary amine, as is diethylamine.

$$H_2NCH_2CH_3 \qquad CH_3CH_2\underset{\underset{H}{|}}{N}CH_2CH_3 \qquad CH_3\underset{\underset{H}{|}}{-}N-CH_2CH_3$$

Ethylamine          Diethylamine          $N$-Methylethylamine
(aminoethane)

*Problem 12–2. Label the following as primary amines, secondary amines or tertiary amines: 2-aminobutane, N-methylpiperidine, N-methylaniline, triethylamine.*

A nitrogen atom carrying four alkyl groups is present in quaternary ammonium ions such as the tetramethylammonium ion.

$$CH_3-\overset{\overset{CH_3}{|}}{\underset{\underset{CH_3}{|}}{N}}^{\oplus}-CH_3 \qquad\qquad CH_3-\overset{\overset{H}{|}}{\underset{\underset{H}{|}}{N}}^{\oplus}-CH_3$$

Tetramethylammonium          Dimethylammonium ion
ion

Ammonium ions with fewer alkyl groups are also possible, for example the dimethylammonium ion, $(CH_3)_2\overset{\oplus}{N}H_2$.

Most simple amines are gases or liquids. Those in which the alkyl group(s) is (are) small are soluble in water. Most small amines have smells like ammonia. Alkylammonium salts are crystalline solids soluble in water and have no smell.

Physical properties of some typical amines

| Formula | Name | B.p. (°C) | Solubility in water |
|---|---|---|---|
| $CH_3NH_2$ | Methylamine | $-7$ | Very soluble |
| $(CH_3)_3N$ | Trimethylamine | 4 | Very soluble |
| $(C_2H_5)_3N$ | Triethylamine | 90 | Slightly soluble |
| $C_5H_{10}NH$ | Piperidine | 106 | Soluble |
| $C_6H_{11}NH_2$ | Cyclohexylamine | 134 | Slightly soluble |
| $C_6H_5NH_2$ | Aniline (phenylamine) | 184 | Insoluble |

The chemistry of alkylammonium ions is comparable with that of alkyloxonium ions, that of tertiary amines with that of ethers and that of primary and secondary amines with that of alcohols.

$$(CH_3)_4N^{\oplus} \quad (CH_3)_3\overset{\oplus}{N}H \quad (CH_3)_3N \quad (CH_3)_2NH \quad CH_3NH_2$$
$$(CH_3)_3O^{\oplus} \quad (CH_3)_2\overset{\oplus}{O}H \quad (CH_3)_2O \qquad\quad CH_3OH$$

## REACTIONS OF THE LONE-PAIR ELECTRONS OF AMINES

All amines have a lone pair of electrons, and the reactions involving this lone pair are the most important aspect of the chemistry of amines.

The lone pair can be used to form a bond to a proton. Amines are moderately strong bases, being stronger bases than water and alcohols but weaker than hydroxide ion. Aqueous solutions of amines are therefore alkaline. The amine is partly protonated in water:

$$RNH_2 + H_2O \ \rightleftharpoons \ R\overset{\oplus}{N}H_3 + {}^{\ominus}OH$$

and is completely converted into an ammonium ion by strong acids:

$$R_2NH + HCl \ \rightarrow \ R_2\overset{\oplus}{N}H_2 \ {}^{\ominus}Cl$$

The salts formed in this reaction are alkylammonium chlorides or amine hydrochlorides. Thus, $CH_3NH_3^{\oplus}$ $Br^{\ominus}$ is called methylammonium bromide or methylamine hydrobromide. These salts are soluble in water since they are ionic. Amines such as dimethylamine are soluble in water owing to hydrogen bonding, but those such as aniline with large hydrocarbon groups are only slightly soluble in water (compare alcohols, p. 95).

Mono-, di-, and trialkylammonium ions are weakly acidic in aqueous solution. They will lose a proton to any strong base and form the amine:

$$(CH_3)_2\overset{\oplus}{N}H_2 + {}^{\ominus}OH \ \rightarrow \ (CH_3)_2NH + H_2O$$

*Problem 12–3. Why are the tetraalkylammonium ions not acidic?*

*Problem 12–4. Suggest a chemical procedure for separating aniline from ethylbenzene. Draw mechanisms for the proton transfer reactions involved.*

The lone pair of electrons of amines can be used to form a bond to carbon. Amines are nucleophilic and react faster than alcohols or water. They react with alkyl halides (and also with carbonyl compounds and acid chlorides, as will be seen in Chapters 15 and 16) with the formation of a new N—C bond:

$$C_2H_5-NH_2 + CH_3-\overline{\underline{Br}}| \ \rightarrow \ C_2H_5-\overset{\overset{\displaystyle H}{|}}{\underset{\underset{\displaystyle H}{|}}{N^{\oplus}}}CH_3 + |\overline{\underline{Br}}|^{\ominus}$$

The product of this reaction is an ammonium ion. Loss of a proton to a base (which could be excess of ethylamine) would then form *N*-methylethylamine:

$$C_2H_5-\overset{\overset{\displaystyle H}{|}}{\underset{\underset{\displaystyle H}{|}}{N^{\oplus}}}CH_3 + {}^{\ominus}OH \ \rightarrow \ C_2H_5-\overset{\overset{\displaystyle H}{|}}{\underset{\underset{\displaystyle H}{|}}{N}}-CH_3 + H_2O$$

The product of this reaction is an ammonium ion. Loss of a proton to a base (which could be excess of ethylamine) would then form *N*-methylethylamine:

## REACTIONS OF THE N–H AND N–C BONDS

Amines can be deprotonated by very strong bases:

$$R_2NH + CH_3Li \rightarrow R_2N^{\ominus} Li^{\oplus} + CH_4$$

whereas ammonium ions with at least one N–H bond can donate protons to water, as we saw above.

Cleavage of the C–N bonds of ammonium ions can occur if the ions can undergo attack at the carbon atom by a nucleophile:

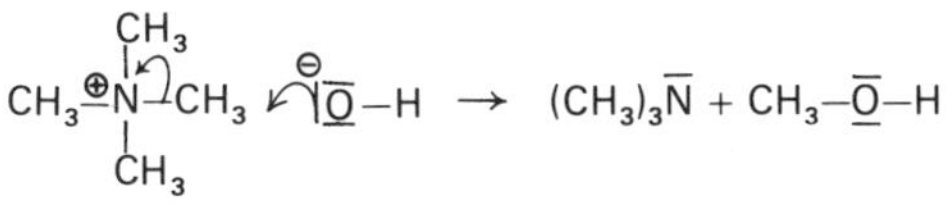

$$CH_3-\overset{\overset{\displaystyle CH_3}{|}}{\underset{\underset{\displaystyle CH_3}{|}}{\overset{\oplus}{N}}}-CH_3 \quad {}^{\ominus}\overline{\underline{O}}-H \rightarrow (CH_3)_3\overline{N} + CH_3-\overline{\underline{O}}-H$$

or attack at a $\beta$-hydrogen atom with elimination of an amine:

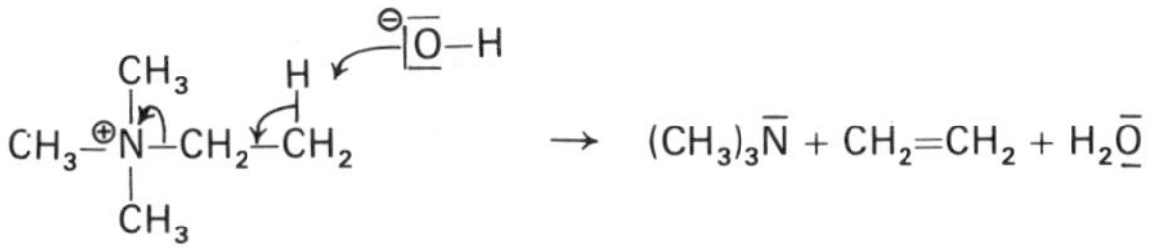

$$CH_3-\overset{\overset{\displaystyle CH_3}{|}}{\underset{\underset{\displaystyle CH_3}{|}}{\overset{\oplus}{N}}}-CH_2-\overset{H}{\underset{}{C}}H_2 \quad {}^{\ominus}\overline{\underline{O}}-H \rightarrow (CH_3)_3\overline{N} + CH_2{=}CH_2 + H_2\overline{\underline{O}}$$

These reactions are direct analogues of $S_N$ and $E$ reactions of alkyl halides and of alkyloxonium ions. They are not observed with trialkylammonium ions because the hydroxide ion would remove a proton from the nitrogen atom. That reaction is faster than the substitution or elimination reactions.

*Problem 12–5. How could propene be made from* NN-*dimethylpropylamine?*

## DIAZONIUM IONS

There is one special group of ammonium ions in which the C–N bond breaks very easily. These are the class of diazonium ions, $R-\overset{\oplus}{N}{\equiv}N$. The syllable 'az' in the word di-az-onium comes from the French word azote, meaning nitrogen.

*Problem 12–6. What will be formed if this ion undergoes simple heterolysis of the* C–N *bond?*

Diazonium ions can be made from primary amines by treatment with nitrous acid and a strong acid such as HCl:

$$R-NH_2 + HONO + HCl \rightarrow R-\overset{\oplus}{N}{\equiv}N \ Cl^{\ominus} + 2H_2O$$

an alkyldiazonium chloride

*Problem 12–7. This conversion is believed to involve the intermediates below:*

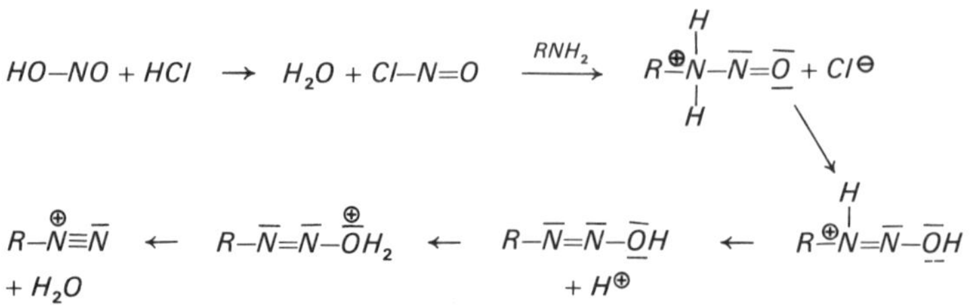

$$HO-NO + HCl \longrightarrow H_2O + Cl-N=O \xrightarrow{RNH_2} R\overset{\oplus}{\underset{H}{\overset{H}{N}}}-\bar{N}=\bar{O} + Cl^{\ominus}$$

$$R-\overset{\oplus}{N}\equiv\bar{N} \longleftarrow R-\bar{N}=\overset{\oplus}{N}-\bar{O}H_2 \longleftarrow R-\bar{N}=\bar{N}-\bar{O}H \longleftarrow R\overset{\oplus}{\underset{H}{\overset{H}{N}}}=\bar{N}-\bar{O}H$$
$$+ H_2O \qquad\qquad\qquad + H^{\oplus}$$

*Check that this mechanism tallies with the overall equation above and draw in all the necessary arrows.*

Heterolysis of the C—N bond occurs below room temperature in the case of diazonium ions made from alkylamines and at rather higher temperatures in the case of benzenediazonium ion made from aniline. In both cases the leaving group is a very stable molecule, $N_2$ gas.

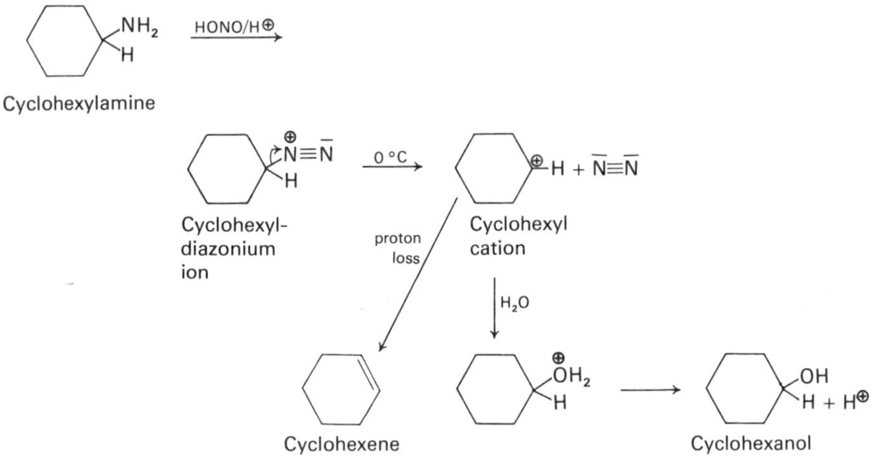

Cyclohexylamine

Cyclohexyl-
diazonium
ion

Cyclohexyl
cation

Cyclohexene

Cyclohexanol

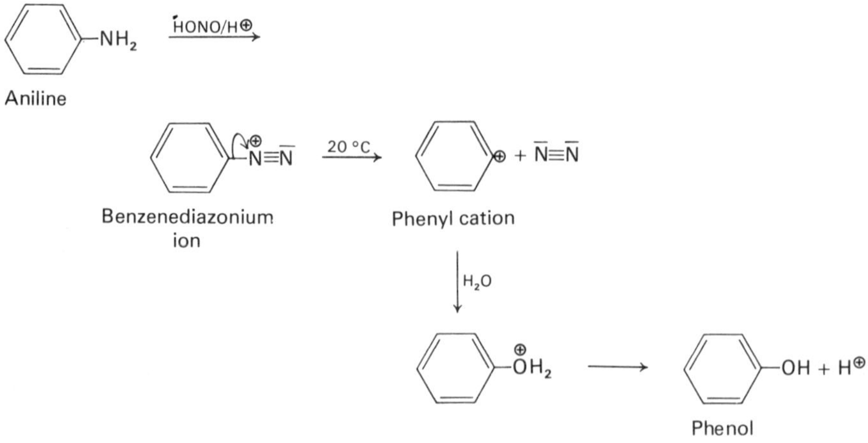

Aniline

Benzenediazonium
ion

Phenyl cation

Phenol

The carbonium ions can combine with any available nucleophile, which might simply be the solvent water. The cyclohexyl cation can also lose a proton to give an alkene.

*Problem 12–8. Draw mechanisms for the formation of cyclohexene and cyclohexanol from cyclohexyldiazonium ion. Are these reactions of types we have seen before?*

*Problem 12–9. What structure would be formed if the phenyl cation lost a proton? Is a molecule with this structure likely to be stable?*

The reaction $R-\overset{\oplus}{N}\equiv N \rightarrow R\overset{\oplus}{O}H_2$ is a substitution which resembles the reaction of tetramethylammonium ion with hydroxide ion (p. 107) or the reaction of ethyl bromide with iodide ion (p. 83). The reaction $R_2CHCR_2-\overset{\oplus}{N}\equiv N \rightarrow R_2C=CR_2 + N_2 + H^\oplus$ is an elimination reaction which resembles the reaction of ethyltrimethyl-ammonium ion with hydroxide ion (p. 107) or the reaction of ethyl bromide with hydroxide ion (p. 86).

By conversion first into diazonium ions, amines can be used to make several other types of compound, just as alkyl halides can, and by similar mechanisms. The substitution and elimination reactions of diazonium ions are exceptionally fast because of the polarity of the $C-N^\oplus$ bond and the stability of the nitrogen molecule which is formed as the bond is broken.

*Problem 12–10. If cyclohexylamine were treated with NaNO₂ + HCl in aqueous methanol, what product would you expect to find together with cyclohexene and cyclohexanol?*

## ANSWERS TO PROBLEMS IN CHAPTER 12

*Problem 12–1.*

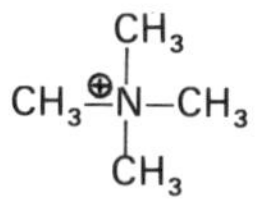

$$CH_3-\overset{\overset{\displaystyle CH_3}{|}}{\underset{\underset{\displaystyle CH_3}{|}}{\overset{\oplus}{N}}}-CH_3$$

and other tetraalkylammonium ions are related to the ammonium ion:

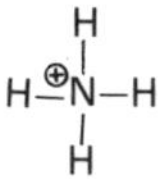

$$H-\overset{\overset{\displaystyle H}{|}}{\underset{\underset{\displaystyle H}{|}}{\overset{\oplus}{N}}}-H$$

*Problem 12–2.*

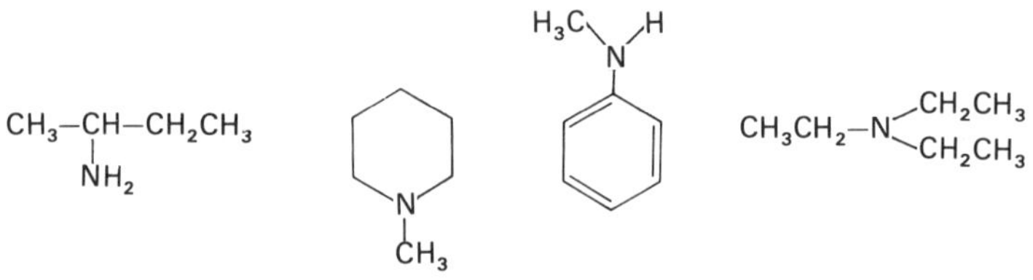

| 2-Aminobutane, | *N*-Methylpiperidine, | *N*-Methyl-aniline, | Triethylamine, |
|---|---|---|---|
| a primary amine | a tertiary amine | a secondary amine | a tertiary amine |

*Problem 12–3.* Tetraalkylammonium ions have no proton on the positive nitrogen.

*Problem 12–4.* Shake the mixture of aniline and ethylbenzene with dilute HCl. The aniline will dissolve into the aqueous layer as anilinium chloride (phenylammonium chloride). The ethylbenzene will not dissolve and can be separated off. On addition of NaOH to the aqueous layer, the aniline will be reformed and will separate out since it is not soluble in water.

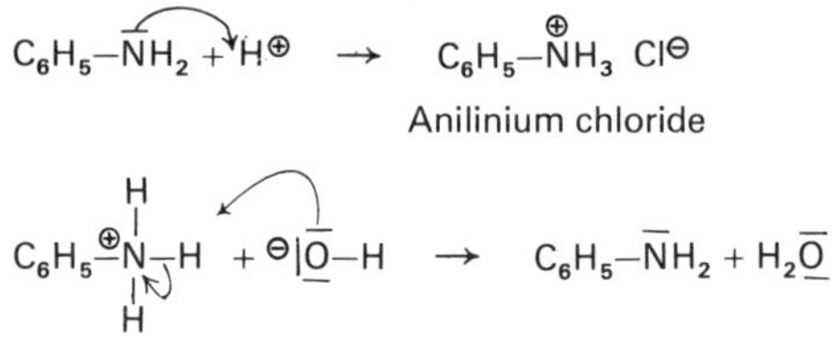

Anilinium chloride

*Problem 12–5.* An elimination reaction is required. This will not happen unless the nitrogen is positive. Therefore, treat the tertiary amine with methyl iodide to give a quaternary ammonium salt. Treat this with NaOH to give water, propene, and trimethylamine. This elimination process is faster than the alternative $S_N$ reactions. Very little propanol or methanol is found among the products.

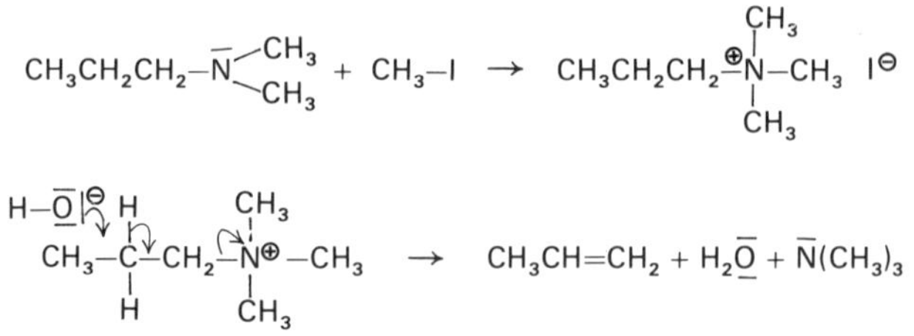

*Problem 12–6.*

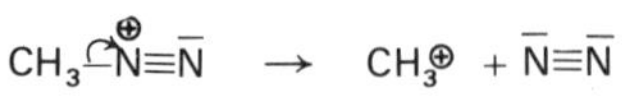

| Methyl cation | Nitrogen gas |

Problem 12–7.

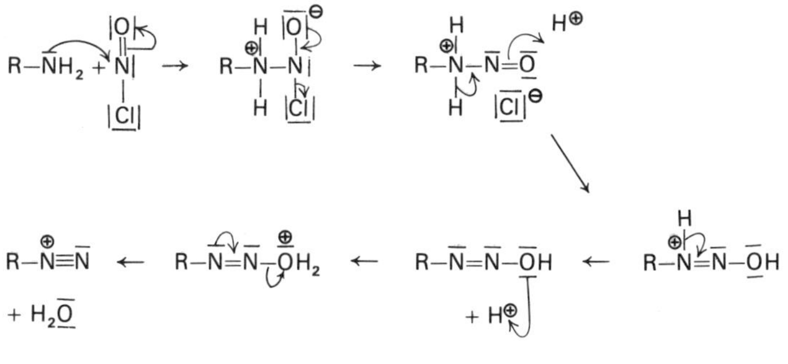

$$R-NH_2 + N\begin{smallmatrix}|O|\\||\\|Cl|\end{smallmatrix} \rightarrow R-N\begin{smallmatrix}H\\|\oplus\\|\\H\end{smallmatrix}-N\begin{smallmatrix}|O|^\ominus\\|\\|Cl|\end{smallmatrix} \rightarrow R-N\begin{smallmatrix}H\\|\\|\\H\end{smallmatrix}-\bar{N}=\bar{O}\quad H^\oplus$$
$$|Cl|^\ominus$$

$$R-N\equiv\bar{N} \leftarrow R-\bar{N}=N-\overset{\ominus}{O}H_2 \leftarrow R-\bar{N}=\bar{N}-\bar{O}H \leftarrow R-N\overset{H}{\overset{\oplus}{=}}\bar{N}-\bar{O}H$$
$$+ H_2\bar{O} \qquad\qquad + H^\oplus$$

Problem 12–8.

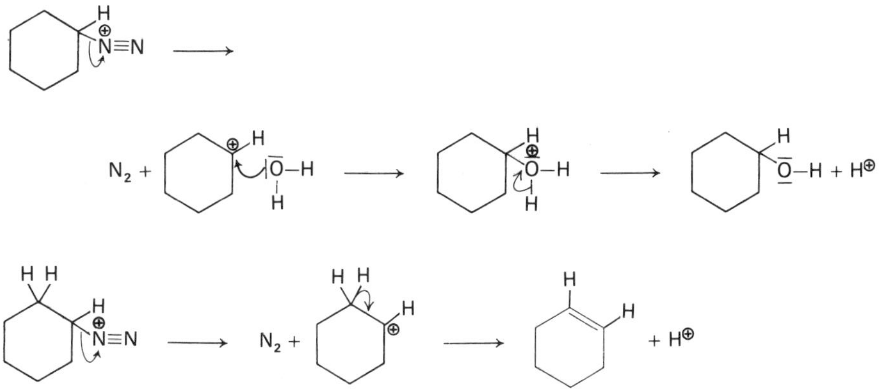

Problem 12–9.

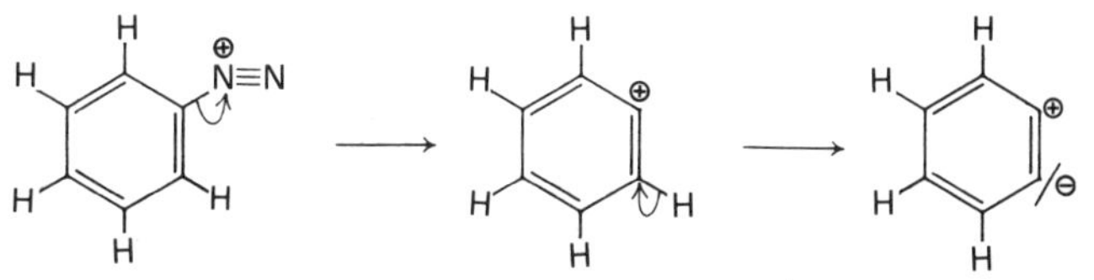

The product would be the highly unstable and reactive molecule shown with opposite charges on adjacent atoms. The formation of a triple bond:

is not possible since acetylenes are linear (see p. 36) and cannot be bent into a ring of this size.

*Problem 12–10.* The diazonium ion would be formed and would break down to the cyclohexyl cation, which could capture methanol or water or lose a proton:

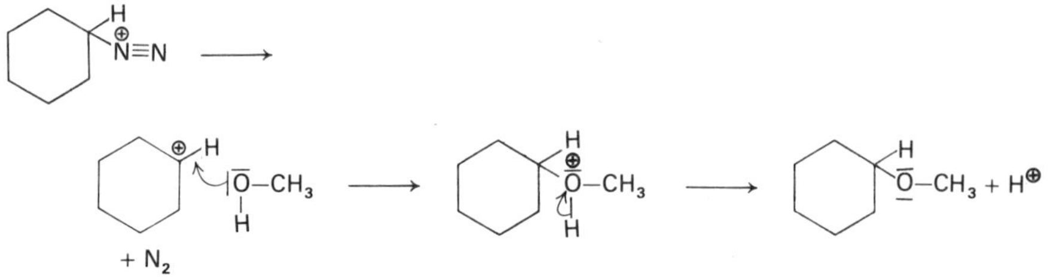

Some cyclohexyl methyl ether is to be expected among the products.

# Chapter 13

## Alkenes

Having studied the behaviour of substances in which carbon is singly bonded to other atoms, we must turn to situations where the carbon is doubly or triply bonded, as in the following types of compounds:

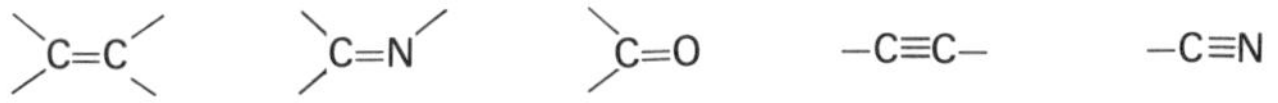

$$\text{\textbackslash C=C/} \qquad \text{\textbackslash C=N/} \qquad \text{\textbackslash C=O} \qquad -C\equiv C- \qquad -C\equiv N$$

| Alkenes | Imines | Ketones | Alkynes | Nitriles |

All of these substances are said to be *unsaturated*, since it is possible in theory (and in practice) to add atoms, such as hydrogen atoms, to them to produce the saturated alkanes, amines, alcohols, etc., discussed above.

*Alkenes* are hydrocarbons containing a carbon–carbon double bond. They are sometimes called *olefins*. Acyclic alkenes with one double bond have the general formula $C_nH_{2n}$. Polyenes (with several double bonds) and cyclic alkenes are, of course, possible. The names include the suffix -ene and the numbers are those of the first carbon (only) of each double bond:

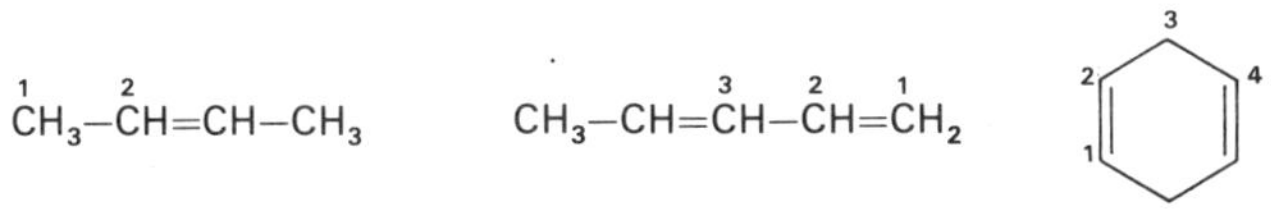

$$\overset{1}{C}H_3-\overset{2}{C}H=CH-CH_3 \qquad\qquad CH_3-\overset{3}{C}H=\overset{2}{C}H-\overset{1}{C}H=CH_2$$

But-2-ene　　　　　　Penta-1,3-diene　　Cyclohexa-1,4-diene

If double bonds and single bonds alternate along the carbon chain so that there is a continuous row of trigonal atoms (see pp. 33–36), then the double bonds are said to be *conjugated*. The double bonds of penta-1,3-diene are conjugated. Those of a 1,4-diene such as cyclohexa-1,4-diene are not conjugated.

Because all of the atoms in alkenes have almost the same electronegativity, these substances are non-polar or very slightly polar and they have physical properties very like those of the alkanes with the same carbon skeletons. They are gases or liquids and are insoluble in water.

113

Boiling points of some alkenes

| Alkene | B.p. (°C) | Alkene | B.p. (°C) |
|---|---|---|---|
| Ethene | −102 | But-1-ene | −7 |
| Propene | −48 | cis-But-2-ene | +4 |
| Cyclohexene | 83 | trans-But-2-ene | +1 |

It is found that the shapes of alkene molecules are such that the two doubly bonded carbon atoms and the four atoms attached directly to them all lie in a plane. Thus, all of the atoms of ethene lie in one plane and all of the carbon atoms of 2,3-dimethylbut-2-ene lie in one plane:

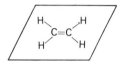

Ethene
(often called ethylene);
all six atoms lie in the
plane of the paper

2,3-Dimethylbut-2-ene;
all six carbon atoms lie
in the plane of the paper

Rotation of the group at one end of the double bond relative to the other is not possible at room temperature and so geometrically isomeric (*cis* and *trans*) forms of some alkenes such as but-2-ene exist (see Chapter 4).

## ADDITION REACTIONS

The main type of reaction of alkenes is the addition of reagents to form saturated molecules. Not all of these additions proceed by the same mechanism.

The addition of buta-1,3-diene to certain substituted ethenes was mentioned in Chapter 7. Each end of the butadiene adds to one end of the ethene. This addition proceeds by a cyclic transition state and is called a pericyclic reaction (see p. 64):

$$H-C\overset{\diagup CH_2}{\underset{\diagdown CH_2}{\big|}} \quad \overset{CHX}{\underset{CH_2}{\|}} \quad \longrightarrow \quad$$

The addition of borane to ethene was mentioned on p. 91.

$$CH_2=CH_2 \quad \longrightarrow \quad \underset{H_2B \quad H}{CH_2-CH_2} \quad \xrightarrow{\text{further additions}} \quad (CH_3CH_2)_3B$$

This addition is also a pericyclic reaction.

A radical addition can be used to make some polymers (addition polymers):

$$R\cdot + CH_2=CH_2 \quad \longrightarrow \quad R-CH_2-\dot{C}H_2 \quad \xrightarrow{CH_2=CH_2} \quad RCH_2CH_2CH_2\dot{C}H_2$$

$$\xrightarrow{\text{further additions}} \quad R-(CH_2CH_2)_nCH_2\dot{C}H_2 \quad \xrightarrow{H-R'} \quad R-(CH_2CH_2)_{n+1}H + \dot{R}'$$

In this process, groups become attached to each end of each ethene unit. The nature and source of the initiating radical R· and of the molecule HR′, which stops the chain extension by transferring a hydrogen atom to the long chain radical, need not concern us here. The chain will be very largely made of repeating $-CH_2CH_2-$ units and the polymer made from ethene (ethylene) is called polyethylene or 'Polythene'. Other alkenes can also be polymerized. Important ones are polypropylene from propene, polystyrene from styrene (phenylethene) and poly(vinyl chloride) (PVC) from vinyl chloride (chloroethene). These polymers have the regular structure

$$\sim\sim\sim CH_2-\underset{\underset{X}{|}}{CH}-CH_2-\underset{\underset{X}{|}}{CH}\sim\sim\sim$$

where X is $CH_3$, $C_6H_5$, and Cl in the three cases given.

## Hydrogenation

An important addition reaction of alkenes is their combination with hydrogen to give alkanes. Overall, one hydrogen atom is added to each end of the double bond:

$$CH_2{=}CH_2 + H_2 \xrightarrow{\text{(Pt)}} CH_3CH_3$$

Alkenes and hydrogen do not react except in the presence of a catalyst. A useful catalyst is platinum (or nickel) metal, powdered or spread as a film with a large surface area. The alkene and the hydrogen become attached to the platinum atoms of the surface. They are adsorbed on to the surface. The formation of the carbon–platinum and hydrogen–platinum bonds weakens or breaks the carbon–carbon and hydrogen–hydrogen bonds of the reagents, making the combination process easier. The precise mechanism is not known. The platinum is a heterogeneous catalyst, since it is a solid while the reagents are gases or liquids or solutions forming a separate phase. Catalytic hydrogenation converts cyclohexene into cyclohexane and both cis- and trans-but-2-ene into butane:

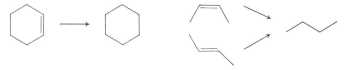

Many vegetable oils contain C=C double bonds. Catalytic hydrogenation of these oils saturates the oils by adding hydrogen to these double bonds. The products are usually solid fats and are sold as margarine.

## REACTIONS OF ALKENES WITH ELECTROPHILES OR NUCLEOPHILES

None of the addition reactions of alkenes described so far have been ionic. If an alkene were to react with an electrophile such as a proton or a carbonium ion, the product would be a carbonium ion:

$$\begin{array}{c}\text{C}=\text{C} \\ \overset{\curvearrowleft}{\text{H}^{\oplus}}\end{array} \longrightarrow \text{H}-\overset{|}{\underset{|}{\text{C}}}-\overset{\oplus}{\text{C}}$$

$$\begin{array}{c}\text{C}=\text{C} \\ \overset{\curvearrowleft}{\text{R}^{\oplus}}\end{array} \longrightarrow \text{R}-\overset{|}{\underset{|}{\text{C}}}-\overset{\oplus}{\text{C}}$$

If an alkene were to react with a nucleophile such as a hydride ion or a carbanion, the product would be a carbanion:

$$\begin{array}{c}\text{C}=\text{C} \\ \overset{\curvearrowright}{\text{H}}{}^{\ominus}\end{array} \longrightarrow \text{H}-\overset{|}{\underset{|}{\text{C}}}-\overset{\ominus}{\text{C}}$$

$$\begin{array}{c}\text{C}=\text{C} \\ \overset{\curvearrowright}{\text{R}}{}^{\ominus}\end{array} \longrightarrow \text{R}-\overset{|}{\underset{|}{\text{C}}}-\overset{\ominus}{\text{C}}$$

Examples of both of these types of process are known. The addition of an electrophile will occur more readily if the substituents on the double bond of the alkene can stabilize the cation which is produced. Likewise, substituents which can stabilize an anionic centre will improve the chances of the second type of reaction occurring.

*Problem 13–1. From your knowledge of delocalization and electronegativity, suggest substituents which would stabilize (a) a carbonium ion and (b) a carbanion.*

### Substituents which favour reactions with electrophiles

Protons add to 1,2-dimethoxyethene to give a cation which is delocalized and therefore more stable than the cation formed by protonation of ethene:

$$H^{\oplus} + CH_3O-CH=CH-OCH_3 \longrightarrow \underset{H}{\overset{H}{CH_3-O-\overset{|}{C}-\overset{\oplus}{C}}}\overset{O-CH_3}{\underset{H}{}} \longleftrightarrow \underset{H}{\overset{H}{CH_3-O-\overset{|}{C}-C}}\overset{\overset{\oplus}{O}-CH_3}{\underset{H}{}}$$

$$H^{\oplus} + CH_2=CH_2 \longrightarrow CH_3-\overset{\oplus}{C}\overset{H}{\underset{H}{}}$$

Any substituents attached to the alkene double bond which are able to release electrons to the carbon atom which is becoming positive during the reaction, and which stabilize the cation by delocalizing its charge, will favour reactions of the alkene with electrophiles. Examples of such substituents include $-OH$, $-OR$, $-NR_2$, and $-O^{\ominus}$.

Protons add to but-2-ene to give a cation which is found to be more stable than the ethyl cation:

$$CH_3-CH=CH-CH_3 + H^\oplus \rightarrow CH_3-CH_2-\overset{\oplus}{C}\overset{CH_3}{\underset{H}{\diagup}}$$

$$CH_2=CH_2 + H^\oplus \rightarrow CH_3-\overset{\oplus}{C}\overset{H}{\underset{H}{\diagup}}$$

No delocalization involving lone pairs of electrons is possible here. The methyl group must be able to reduce the electron deficiency at the positive carbon in some other way. The origin of this effect is not completely clear. We can perhaps account for the fact that the alkyl groups attached to positively charged carbon atoms stabilize the cation by assuming that the carbon of the alkyl group is less electronegative than hydrogen.

$$^\oplus CH_3 < CH_3-\overset{\oplus}{C}H_2 < CH_3-\overset{\oplus}{C}H-CH_3 < CH_3-\overset{\oplus}{C}\overset{CH_3}{\underset{CH_3}{\diagup}}$$

Some carbonium ions in order of increasing stability

Thus, alkenes in which the double bond carries alkyl groups or hydroxyl, alkoxyl, or amino groups will be better electron donors than ethene.

### Substituents which favour reactions with nucleophiles

Alkenes in which the double bond carries fluorine atoms or cyano, nitro, or carbonyl groups will behave as electrophiles more readily than does ethene. Thus, ethene does not react with nucleophilic reagents whereas tetrafluoroethene and cyanoethene can. Tetrafluoroethane combines with dimethylamine:

$$F_2C=CF_2 + HN(CH_3)_2 \rightarrow (CH_3)_2\overset{\overset{H}{\underset{|}{\oplus}}}{N}-CF_2-\overset{\ominus}{C}F_2 \rightarrow (CH_3)_2N-CF_2-CF_2-H$$

The negative charge on the carbon in the intermediate is stabilized by the electron-withdrawing effect of the strongly electronegative fluorine atoms. When cyanoethene (acrylonitrile) reacts with methoxide ion, the carbanion formed is a delocalized one in which the negative charge is partly carried by the nitrogen:

$$CH_3-\overset{|}{\underset{|}{O}}|^\ominus + CH_2=C\overset{C\equiv\overline{N}}{\underset{H}{\diagup}} \rightarrow CH_3-\overset{-}{O}-CH_2-\overset{\ominus}{C}\overset{C\equiv\overline{N}}{\underset{H}{\diagup}}$$

$$\updownarrow$$

$$CH_3-\overset{-}{O}-CH_2-C\overset{C=\overline{N}|^\ominus}{\underset{H}{\diagup}}$$

### Conclusions

We can conclude that those alkenes which carry substituents that can stabilize a carbonium ion will combine with electrophiles, and those alkenes which carry substituents that can stabilize a carbanion will combine with nucleophiles.

## UNSYMMETRICAL ALKENES

So far we have not been concerned about which end of the double bond becomes bonded to the reagent. It seems from the last example above that the effects of substituents which determine the reactivity of an alkene compared to ethene can also determine the relative reactivity of the ends of an unsymmetrical alkene when it reacts with a nucleophile or an electrophile.

The arguments used above can be applied to explain the following observations.

(a) $CH_3-CH=CH_2 + H^\oplus$

$\longrightarrow CH_3-\overset{\oplus}{CH}-CH_3$ (A)

$\overset{\times}{\longrightarrow} CH_3-CH_2-\overset{\oplus}{CH_2}$ (B)

It happens that, in this reaction, product selection is under thermodynamic control. The cation (A) which is formed has two alkyl groups attached to the positive carbon and so is more stable than the alternative product (B) which has only one.

(b) Addition of the $HOO^\ominus$ ion, made from hydrogen peroxide and a base, to the C=C double bond of cyclohex-2-enone produces preferentially the anion (C). It happens that in this reaction, product selection is under kinetic control. The transition state leading to C is of lower energy than that leading to D since there is considerable delocalization in the former transition state but not in the latter (p. 59).

Cyclohex-2-en-1-one

*Problem 13–2.*

(a) *What ion will be formed preferentially by addition of a proton to 2-methylpropene?*

(b) *What ion will be formed preferentially by addition of a proton to buta-1,3-diene? How many different products are there in this case? Are any of them delocalized?*

(c) *What ion will be formed preferentially by addition of a proton to penta-1,3-diene?*

(d) *What ion will be formed preferentially by addition of cyanide ion to the* $C=C$ *of* $CH_2=CH-C\overset{\nearrow H}{\underset{\searrow O}{}}$ ?

(e) *What ion will be formed preferentially by addition of a proton to* $CH_2=CH-C\overset{\nearrow H}{\underset{\searrow O}{}}$?

# ADDITION REACTIONS IN WHICH THE ALKENE BEHAVES AS A NUCLEOPHILE

We have seen that certain alkenes can behave as nucleophiles and donate electrons to other atoms. Simple alkenes are relatively poor nucleophiles and react with, say, methyl iodide much more slowly than do water or amines. Most alkenes are weaker bases than water, and can be protonated only in strongly acidic media.

If the alkene does act as a nucleophile, it will be converted into a carbonium ion. Addition of protons or other electrophiles to alkenes is a useful method of making certain carbonium ions. The carbonium ions will be strongly electrophilic and usually will capture any nucleophile present to give a saturated molecule, except when the nucleophile is basic. If the nucleophile is basic (e.g. $NH_3$, $CN^{\ominus}$) it will capture a proton from the carbonium ion and re-form an alkene:

$$CH_3-\overset{\oplus}{C}H_2 + {}^{\ominus}CN \rightarrow CH_2=CH_2 + HCN$$

If the nucleophile is not basic, it will add to the carbonium ion.

## Addition of H—Br

When ethene is treated with HBr, the ethene reacts first with the proton to give an ethyl cation, which then combines with the bromide ion (which is not basic) to give the product, bromoethane. The overall effect is addition of HBr to the double bond:

$$\overset{H^{\oplus}}{CH_2{=}CH_2} \longrightarrow CH_3-\overset{\oplus}{C}H_2 \xrightarrow{\boxed{Br}^{\ominus}} CH_3-CH_2-\overline{Br}|$$

Propene reacts in the same way. The proton adds first to give the more stable of the two possible cations. This cation then captures bromide ion, giving 2-bromopropane:

$$CH_3-CH{=}CH_2 + H^{\oplus} \longrightarrow CH_3-\overset{\oplus}{\underset{H}{C}}-CH_3 \xrightarrow{Br^{\ominus}} CH_3-\overset{Br}{\underset{H}{C}}-CH_3$$

Overall:

$$CH_3-CH{=}CH_2 + HBr \rightarrow CH_3CHBrCH_3$$

The structure of the final product is determined in the first step by the relative stabilities of the two possible intermediate carbonium ions.

## Addition of H—OR

If an alkene is treated with an acid which has a non-nucleophilic anion (e.g. $H_2SO_4$), the intermediate cation will capture the best nucleophile available, which may be a solvent molecule, for example alcohol or water.

$$CH_2=CH-OCH_3 \xrightarrow{H_2SO_4 \text{ in } CH_3OH} CH_3-CH{\overset{\displaystyle OCH_3}{\underset{\displaystyle OCH_3}{<}}}$$

$$\overset{H_{\cdot}^{\oplus}}{CH_2{\overset{\curvearrowleft}{=}}CH-OCH_3} \rightarrow CH_3-\overset{\oplus}{C}H-O-CH_3$$

(Delocalized and much more stable than the alternative isomer)

Then

$$\underset{\displaystyle H-\underset{|}{\overset{\ominus}{O}}-CH_3}{CH_3-\overset{\oplus}{C}H-OCH_3} \rightarrow CH_3-CH{\overset{\displaystyle OCH_3}{\underset{\displaystyle \underset{H}{\overset{|}{\underset{\phantom{.}}{\oplus}}O}-CH_3}{<}}} \rightarrow CH_3CH(OCH_3)_2 + H^{\oplus}$$

The sulphuric acid is a homogeneous catalyst for the addition. It only acts as a source of protons. The overall effect is the addition of methanol (in this case) to the alkene. Again, the structure of the final product is determined in the first step by the relative stabilities of the two possible cations.

### Addition of H—OH

The addition of water to alkenes to give alcohols is called *hydration*. The hydration of simple alkenes such as ethene and propene, which are readily available from petroleum, is an important way of manufacturing alcohols:

$$CH_3-CH=CH_2 \xrightarrow{H^{\oplus} \text{ in } H_2O} \underset{\displaystyle \underset{OH}{\overset{|}{\phantom{.}}}}{CH_3-CH-CH_3}$$

Propan-2-ol

The mechanism is the same as in the case of the addition of methanol to methoxyethene. An acid such as $H_2SO_4$ is used as a catalyst and again the structure of the final product is determined in the first step of the addition by the relative stabilities of the possible cations.

### Addition of bromine

In all of the cases above, the alkene has reacted with a proton in the first step of the addition. Alkenes will react with other electrophiles such as carbonium ions and bromine. The reaction with bromine results in the addition of the bromine molecule to the alkene. The disappearance of the colour when bromine is shaken with an organic compound is a useful indication of the presence of a C=C double bond.

$$\overset{\displaystyle}{C}{=}C + Br_2 \rightarrow \underset{\displaystyle Br\ Br}{C{-}C}$$

The electrophile is the bromine molecule, which is attacked by the alkene in a manner which can be thought of as an $S_N$ reaction at a bromine atom:

$$\underset{\underset{|\overline{Br}\overset{\frown}{-}Br|}{}}{\overset{}{\underset{}{}}C{=}C}\overset{}{\underset{}{}} \quad \rightarrow \quad \underset{\underset{|\overline{Br}|}{}}{\overset{\oplus}{}}C{-}C\overset{}{\underset{}{}} \quad + \quad |\overline{\underline{Br}}|^{\ominus}$$

The bromocarbonium ion so formed might combine with the bromide ion to give the observed product directly, but there is evidence that this does not happen. The initial carbonium ion has a built-in nucleophile and can cyclize to give a delocalized ion in which the charge is carried partly by the bromine:

$$\underset{\underset{|\overline{Br}|}{}}{\overset{\oplus}{}}C{-}C \longrightarrow \underset{\underset{\overline{Br}^{\oplus}}{}}{}C{-}C \longleftrightarrow \underset{\underset{|\overline{Br}|}{}}{\overset{\oplus}{}}C{-}C \longleftrightarrow \underset{\underset{|\overline{Br}|}{}}{\overset{\oplus}{}}C{-}C$$

Cyclic intermediate

This cyclic intermediate can be attacked by the bromide ion to yield the dibromoalkane, which is the product:

$$\underset{\underset{\oplus\overline{Br}}{}}{}C{-}C \overset{|\overline{\underline{Br}}|^{\ominus}}{} \quad \rightarrow \quad \underset{\underset{|\overline{Br}|}{}}{\overset{|\overline{Br}|}{}}C{-}C$$

## ADDITION REACTIONS IN WHICH THE ALKENE BEHAVES AS AN ELECTROPHILE

We have seen that certain alkenes can combine with certain nucleophiles, $X^{\ominus}$, to form carbanions:

$$R_2C{=}CR_2 + X^{\ominus} \quad \rightarrow \quad R_2C{-}\overset{\ominus}{C}R_2$$
$$\underset{X}{|}$$

This provides a useful synthesis of certain carbanions. The carbanions will be able to react as nucleophiles, forming bonds to carbon or hydrogen. We shall discuss only the latter possibility here. The carbanions are usually treated with an acid from which they capture a proton:

$$R_2C{-}\overset{\ominus}{C}R_2 + H^{\oplus} \quad \rightarrow \quad R_2C{-}CR_2$$
$$\underset{X}{|} \qquad\qquad\qquad \underset{X\ H}{|\ \ |}$$

The overall effect is the addition of HX to the alkene.

For example, if cyclohex-2-enone is treated with cyanide ion in the presence of a weak acid such as ammonium ion, the overall reaction is addition of HCN to the C=C double bond:

Cyclohex-2-enone                       3-Cyanocyclohexanone

The site of addition of the cyanide ion is determined in the first step (see pp. 59 and 117).

## CONCLUSION

We conclude that alkenes readily undergo various kinds of addition reactions. The mechanisms may involve pericyclic processes, radicals, heterogeneous catalysts, or ions. Ionic addition can proceed via an intermediate carbonium ion or an intermediate carbanion depending on the substituents on the alkene double bond.

*Problem 13–3. Suggest how cyclopentene could be converted into:*
  (a) *cyclopentane;*

  (b) *triscyclopentylborane,* ;

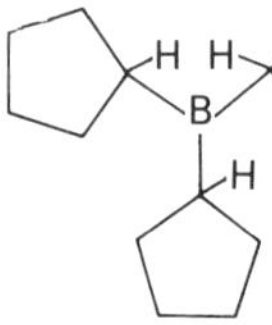

  (c) *cyclopentanol;*
  (d) *1,2-dibromocyclopentane;*
  (e) *cyanocyclopentane.*

*Problem 13–4. What products would result from the following reactions?:*

  (f)  $CH_2=C{<}^{CH_3}_{CH_3}$  $\xrightarrow[(2)\,H_2O]{(1)\,H^{\oplus}}$

  (g)  $CH_2=CH–CN$  $\xrightarrow[(2)\,H^{\oplus}]{(1)\,^{\ominus}CN}$

*Problem 13–5. Draw a mechanism for the following reaction, which involves addition of a nucleophile and an electrophile to each of the double bonds:*

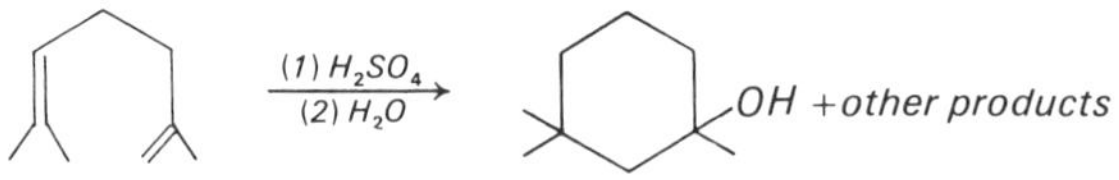

## ANSWERS TO PROBLEMS IN CHAPTER 13

*Problem 13–1.* (a) A carbonium ion $R\overset{\oplus}{C}H_2$ will be more stable than $^{\oplus}CH_3$ if the group R is less electronegative than H or if its presence results in delocalization. We saw in Chapter 6 that the ion $^{\oplus}CH_2–OH$ is much more stable than $^{\oplus}CH_3$ because it is delocalized: $^{\oplus}CH_2–\overline{\underset{\oplus}{O}}–H \leftrightarrow CH_2=\overset{\ominus}{O}–H$. The same will hold for $^{\oplus}CH_2OR$, $^{\oplus}CH_2NR_2$, and $^{\oplus}CH_2–CH=CH_2$. Substituents such as $–OH$, $–OR$, $–NR_2$, and $–CR=CR_2^{\ominus}$ stabilize carbonium ions.

  (b) A carbanion $RCH_2$ will be more stable than $^{\ominus}CH_3$ if the group R is more electronegative than H, e.g. $^{\ominus}CH_2F$ is more stable than $^{\ominus}CH_3$, or if R has a positive charge which increases its pull on the bonding electrons, e.g. $^{\ominus}CH_2-\overset{\oplus}{N}(CH_3)_3$ is more stable than $^{\ominus}CH_3$, or if the presence of R results in delocalization, e.g. the ions

$$\overset{\ominus}{C}H_2-C\overset{\nearrow O}{\underset{\searrow H}{}} \quad \longleftrightarrow \quad CH_2=C\overset{\nearrow O^{\ominus}}{\underset{\searrow H}{}}$$

and

$$^{\ominus}\overline{C}H_2-C\equiv N \quad \longleftrightarrow \quad CH_2=C=\overline{N}^{\ominus}$$

are both more stable than $^{\ominus}\overline{C}H_3$.

*Problem 13–2.*

(a) $CH_2=C\overset{\nearrow CH_3}{\underset{\searrow CH_3}{}} \quad \xrightarrow{H^{\oplus}} \quad CH_3-\overset{\oplus}{C}\overset{\nearrow CH_3}{\underset{\searrow CH_3}{}} \quad$ not $\quad ^{\oplus}CH_2-\overset{\overset{\displaystyle H}{|}}{\underset{\underset{\displaystyle CH_3}{|}}{C}}-CH_3$

since the former has more stabilizing alkyl groups attached to the positive carbon.

(b) $CH_2=CH-CH=CH_2 \quad \xrightarrow{H^{\oplus}} \quad CH_3-\overset{\oplus}{C}H-CH=CH_2 \quad$ not $\quad ^{\oplus}CH_2-CH_2-CH=CH_2$

There are only two possibilities. Only the first one is delocalized see p. 58.

(c) $CH_3-CH=CH-CH=CH_2 + H^{\oplus}$

might give:

$$CH_3-CH=CH-CH_2-\overset{\oplus}{C}H_2 \qquad \text{(A)}$$

or

$$CH_3-CH=CH-\overset{\oplus}{C}H-CH_3 \quad \longleftrightarrow \quad CH_3-\overset{\oplus}{C}H-CH=CH-CH_3 \qquad \text{(B)}$$

or

$$CH_3-CH_2-\overset{\oplus}{C}H-CH=CH_2 \quad \longleftrightarrow \quad CH_3-CH_2-CH=CH-\overset{\oplus}{C}H_2 \qquad \text{(C)}$$

or

$$CH_3-\overset{\oplus}{C}H-CH_2-CH=CH_2 \qquad \text{(D)}$$

B and C are delocalized ions; A and D are not. B has two alkyl groups attached to carbon atoms which carry part of the charge; C has only one. Therefore, B should be the most stable of the four. B is the major product.

(d) $CH_2=CH-C\overset{\nearrow O}{\underset{\searrow H}{}} \quad \xrightarrow{^{\ominus}CN} \quad CH_2-\overset{\ominus}{C}H-C\overset{\nearrow O}{\underset{\searrow H}{}} \quad$ not $\quad ^{\ominus}CH_2-CH-C\overset{\nearrow O}{\underset{\searrow H}{}}$

(with $\overset{|}{\underset{\|}{C}}$ / $\underset{N}{}$ groups below)

since only the first ion is delocalized.

(e) $CH_2{=}CH{-}CH{=}O$ $\xrightarrow{H^{\oplus}}$ $CH_3{-}\overset{\oplus}{C}H{-}CH{=}O$

or

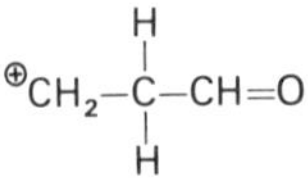

$$\overset{\oplus}{C}H_2{-}\overset{\overset{\displaystyle H}{|}}{\underset{\underset{\displaystyle H}{|}}{C}}{-}CH{=}O$$

or

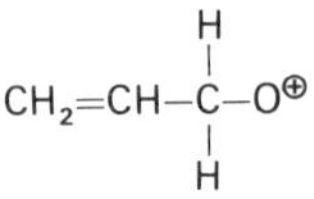

$$CH_2{=}CH{-}\overset{\overset{\displaystyle H}{|}}{\underset{\underset{\displaystyle H}{|}}{C}}{-}O^{\oplus}$$

or

$CH_2{=}CH{-}\overset{\oplus}{C}H{-}\overline{O}{-}H$ $\longleftrightarrow$ $\overset{\oplus}{C}H_2{-}CH{=}CH{-}\overline{O}{-}H$

$\longleftrightarrow$ $CH_2{=}CH{-}CH{-}\overset{\oplus}{O}{-}H$

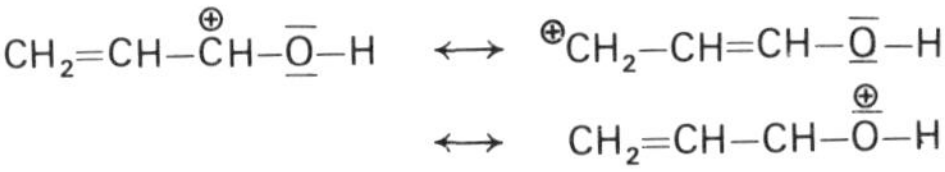

There are four possible products. The last one is extensively delocalized and is by far the most stable, and is the only one formed. See p. 71 and answer (c) p. 123.

*Problem 13–3.*

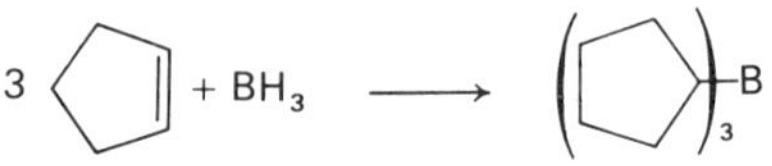

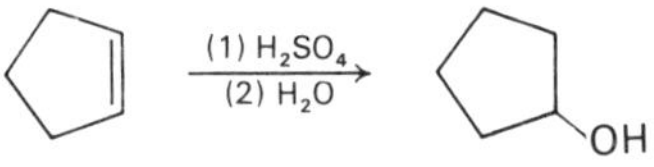

(a) ⬠ + H₂ + Pt catalyst ⟶ ⬠

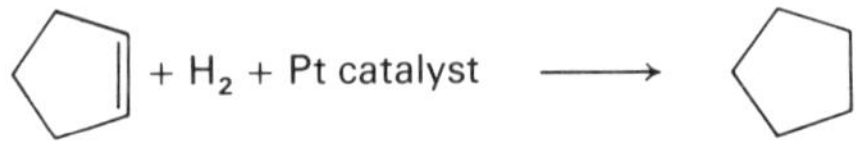

(b) 3 ⬠ + BH₃ ⟶ (⬠)₃B

(c) ⬠ $\xrightarrow[\text{(2) H}_2\text{O}]{\text{(1) H}_2\text{SO}_4}$ ⬠OH

(d) ⬠ + Br₂ ⟶ ⬠(Br, Br)

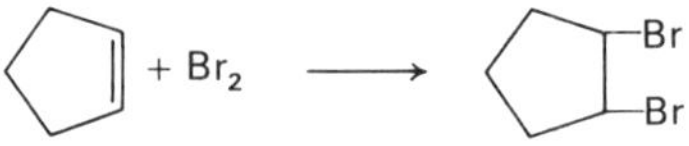

Actually only *trans*-1,2-dibromocyclopentane is formed.

(e) ⬠ $\xrightarrow{HBr}$ ⬠Br $\xrightarrow{^{\ominus}CN}$ ⬠CN + Br^⊖

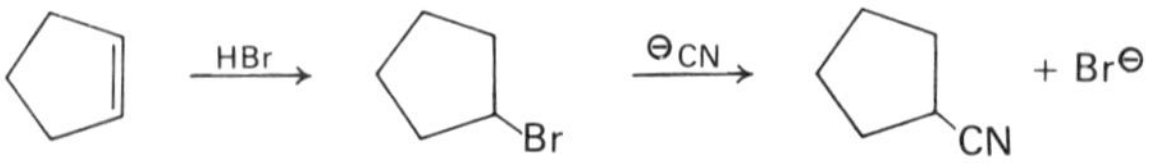

HCN cannot be made to add to alkenes unless the substituents stabilize carbanions. Cyanide ion will not add to cyclopentene since the carbanion is not stable:

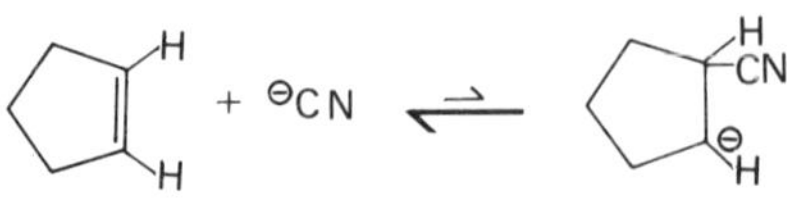

⬠(H, H) + $^{\ominus}CN$ ⇌ ⬠(H–CN, ⊖–H)

If $H_2SO_4$ were used to make cyclopentyl cation and cyanide ion then added, the basic cyanide would remove a proton again to re-form cyclopentene.

*Problem 13–4.*

(*f*) $CH_2=C{\overset{CH_3}{\underset{CH_3}{\diagdown}}}$ $\xrightarrow{\text{H}^\oplus}$ $CH_3-\overset{\oplus}{C}{\overset{CH_3}{\underset{CH_3}{\diagdown}}}$ preferentially.

Addition of water $\longrightarrow$ $(CH_3)_3C-\overset{\oplus}{O}H_2$ $\longrightarrow$ $(CH_3)_3C-OH + H^\oplus$

(*g*) $CH_2=CH-C{\equiv}N$ $\xrightarrow{\ominus\text{CN}}$ $N{\equiv}C-CH_2-\overset{\ominus}{C}H-C{\equiv}N$ preferentially.

Addition of $H^\oplus$ $\longrightarrow$ $NC-CH_2-CH_2-CN$

*Problem 13–5.*

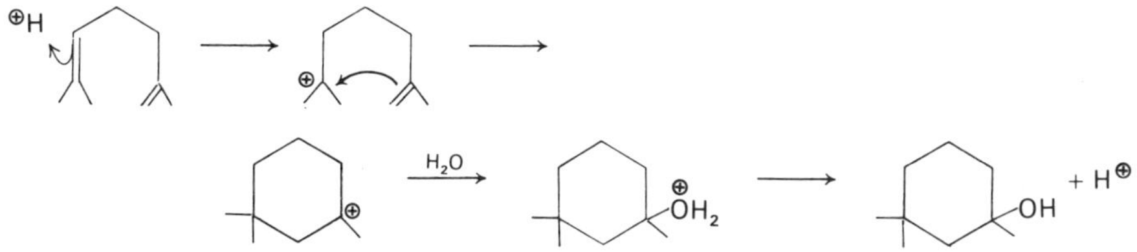

The double bond with most alkyl groups attached is protonated to give the more stable possible cation. The best nucleophile available is the other double bond. Formation of the C—C bond produces a six-membered rather than a five-membered ring in order to form the more stable of the two possible product ions. This combines with the water. The process involves addition of $H^\oplus$ and carbon to the first double bond and of $C^\oplus$ and water to the second.

# Chapter 14

## Benzene and its Derivatives

Benzene is a liquid which boils at 80 °C and freezes at 5.4 °C. It is cheap and has been used as a convenient solvent, but repeated exposure to the vapour even at low concentrations is hazardous and it should be used only in a fume cupboard. Benzene can cause leukaemia, and some compounds containing several benzene rings cause other forms of cancer. Phenol (hydroxybenzene), once called carbolic acid, is toxic orally and causes nasty skin burns. Nitrobenzene and aminobenzene (aniline) are also toxic, whereas other benzene derivatives such as methoxybenzene (anisole), methylbenzene (toluene), and benzoic acid are not considered to be harmful. Benzene and alkylbenzenes, being hydrocarbons are insoluble in water and most are liquids with fairly high boiling points. The physical properties of benzenes carrying functional groups are largely determined by the polarity or hydrogen bonding of these groups.

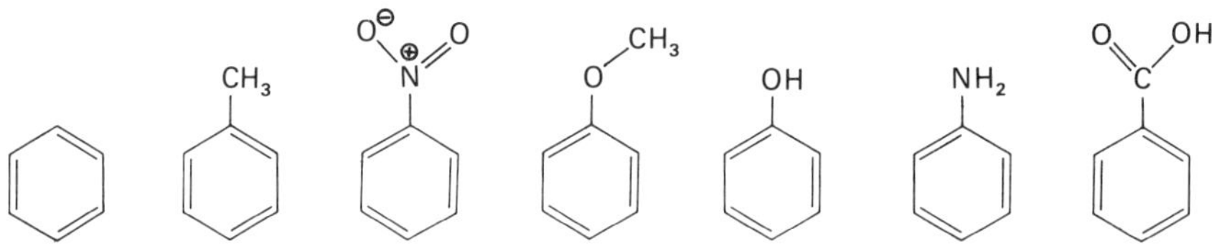

| Benzene | Toluene | Nitrobenzene | Anisole | Phenol | Aniline | Benzoic acid |
|---------|---------|--------------|---------|--------|---------|--------------|
| b.p. 80 °C | b.p. 111 °C | b.p. 210 °C | b.p. 154 °C | m.p. 43 °C | (phenylamine) b.p. 184 °C | m.p. 122 °C |

Benzene has the formula $C_6H_6$, and the molecule is planar and exactly hexagonal. The C—C bonds are all the same length. The carbon atoms are, therefore, trigonal and form three sigma bonds, one to each neighbouring carbon and one to hydrogen. The remaining six electrons are spread uniformly round the ring. Two oversimple pictures can be drawn and the true symmetrical electron distribution can be represented by both of these together as follows:

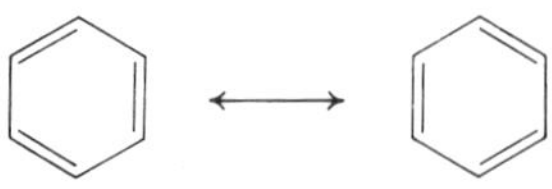

Remember that each corner of the hexagon represents a carbon atom and each of these carries one hydrogen atom. In this chapter we shall use either one of these pictures by itself to represent benzene. We shall not forget that it is an inadequate representation. Benzene can also be represented as

where a new symbol, the circle, represents the six electrons which are delocalized.

Because of the delocalization of these electrons, which is possible because of the cyclic and planar nature of the molecule, benzene is much more stable than isomers such as fulvene and bicyclohexadiene:

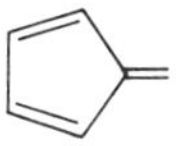

Fulvene 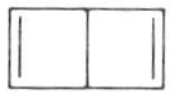       Bicyclohexadiene

The stability of the arrangement of six trigonal atoms in a ring and six electrons delocalized over them is so great that reactions which destroy this system require the input of considerable energy and reactions which produce this system are thermodynamically favoured. Molecules containing this kind of ring are often described as *aromatic*. What has scent to do with stability? When the structures of natural substances were first being determined, it was found that several pleasant smelling compounds contained benzene rings. These compounds were described as aromatic. In due course, all benzenoid compounds were called aromatic whether they had an aroma or not. Later, the meaning of the word was modified again to mean having a planar ring of trigonal atoms with enough electrons delocalized over them to give a stabilized system. This chapter deals only with benzenoid compounds, that is, compounds containing a benzene ring.

All kinds of substituents can be attached to the ring, as the list of compounds opposite suggests. Some further examples are:

Benzyl alcohol (phenylmethanol)    Benzaldehyde (phenylmethanal)    Benzenesulphonic acid    Acetophenone (methyl phenyl ketone; 1-phenylethanone)

If there are two substituents on the ring they can be sited *ortho* to one another, *meta* to one another, or *para* to one another:

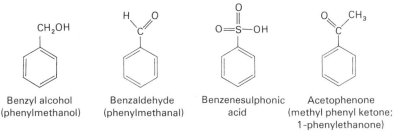

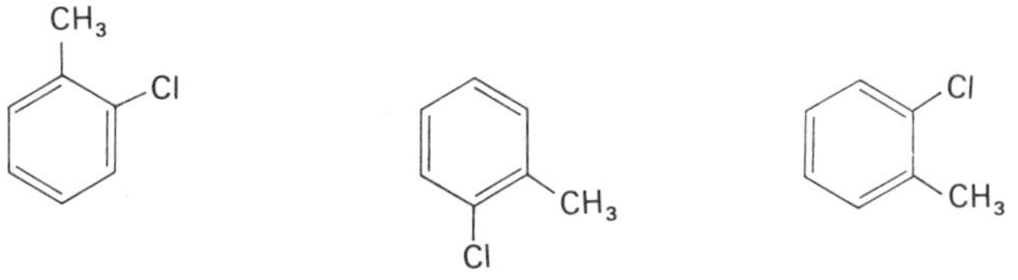

These are all pictures of *ortho*-chlorotoluene

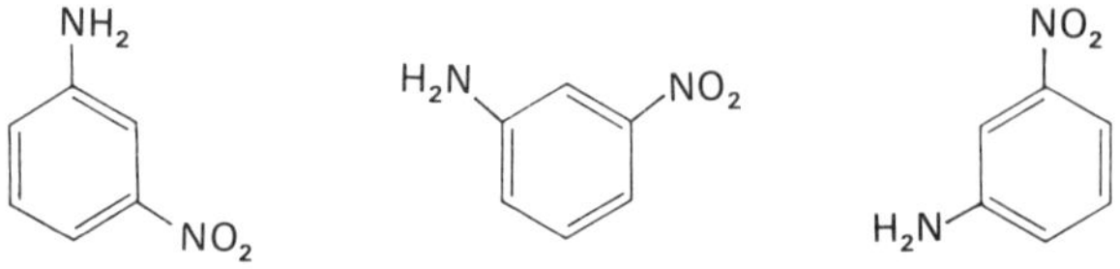

These are all pictures of *meta*-nitroaniline

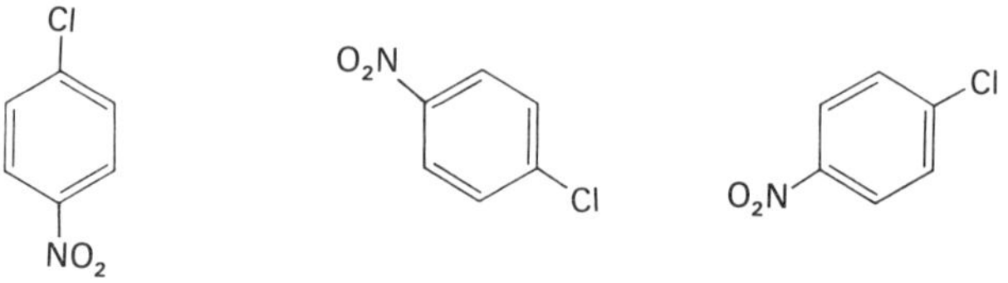

These are all pictures of *para*-chloronitrobenzene

The prefixes *ortho*-, *meta*-, and *para*- are usually contracted to *o*-, *m*- and *p*- in the names of compounds:

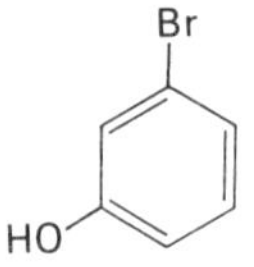

*m*-Bromophenol
or 3-bromophenol

*p*-Dihydroxybenzene or 1,4-dihydroxy-
benzene (usually called hydroquinone)

Normal numbering can also be used as in the cases above and *must* be used for polysubstituted benzenes such as those below. If the compound is named as a substituted benzene the numbering of the ring positions can start at any substituted carbon atom, e.g. 4-chloronitrobenzene or 1,3-dichlorobenzene:

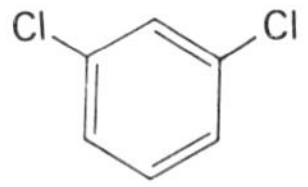

If the compound is named as a substituted toluene or substituted phenol so that the presence of one of the substituents is implied by a special name, then the numbering must start at *that* substituent, e.g. 3-bromophenol and other examples below:

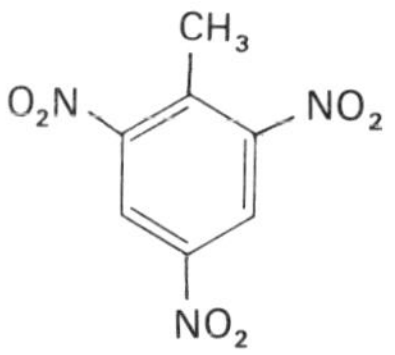

2,4,6-Trinitrotoluene (TNT)

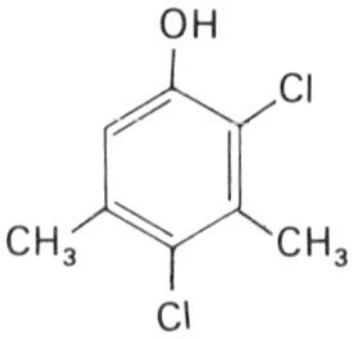

2,4-Dichloro-3,5-dimethylphenol
('Dettol')

Sometimes the benzene ring is named as a substituent attached to something else. The group $C_6H_5$- is called a *phenyl* group. Phenyl groups and substituted phenyl groups are collectively called *aryl* groups. Where do you think the syllable ar- comes from? The group $C_6H_5-CH_2-$ is called a benzyl group and is present in benzyl alcohol (see p. 127).

2,4-Dinitrophenylhydrazine
(hydrazine is $NH_2NH_2$)

*p*-Chlorobenzyl bromide

## REACTIONS OF BENZENE AND SUBSTITUTED BENZENES

The following discussion is divided into three sections which answer the questions:
  (1) do functional groups behave in the same ways when attached to aryl groups as when attached to alkyl groups?
  (2) does the benzene ring itself behave like an alkene or does it undergo new types of reaction? and
  (3) how is the behaviour of the ring affected by substituents?

### 1. Reactions of functional groups

In earlier chapters we discussed functional groups such as $-OH$ attached to alkyl groups and differentiated reactions of the $C-O$ bond, of the $O-H$ bond, and of the lone pairs of electrons. If the $-OH$ group is attached to a trigonal instead of a tetrahedral carbon atom we can guess that reactions of the $C-O$ bond will be different. Elimination reactions cannot occur since it is not possible to build a linear $C-C\equiv C-C$ group into a six-membered ring. $S_N$ reactions also do not occur.

The reactions of the lone pairs of electrons and of the $O-H$ bond should not be much affected when the $-OH$ group is attached to a benzene ring, except in that delocalization may be possible in the reagents or products.

For example, in phenol the lone pairs of electrons on the oxygen atom are delocalized. We can draw several electron distribution pictures which indicate, when taken together, that the $C-O$ bond in phenol is a little like a double bond and the oxygen atom is slightly more deficient in electrons and the ring slightly richer in electrons than the first two pictures on p. 130 suggest.

Such delocalization may reduce the energies of reagents or products and alter the positions of equilibria from those in substituted alkanes. Delocalization may also alter the energies of transition states and so affect the rates of reactions. Various substituents on benzene rings are now discussed in turn.

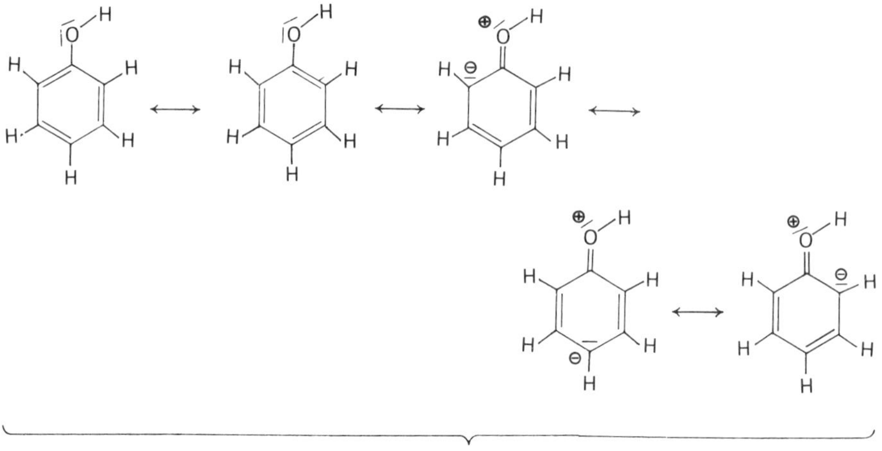

Phenol

## Halogen atoms

In Chapter 9, the reactions of alkyl halides were discussed. Substitution, elimination, and reduction were all possible and all involved production of halide ion. Will any or all of these be possible for aryl halides such as bromobenzene? The main difference lies in the trigonal stereochemistry of the carbon atom to which the halogen is attached. Attack by nucleophiles at such a carbon atom with displacement of bromide ion cannot proceed by the same mechanism as for alkyl halides. Similarly, elimination of bromide ion and a proton, although possible, cannot be analogous to the elimination from bromoethane to give ethene. Substitutions and eliminations are known for bromobenzene. They proceed by different mechanisms from the reactions of bromoethane and they occur only under much more vigorous conditions. Bromobenzene and other aryl halides are rather inert towards bases and nucleophiles. Aryl halides do react with lithium and magnesium in the expected way.

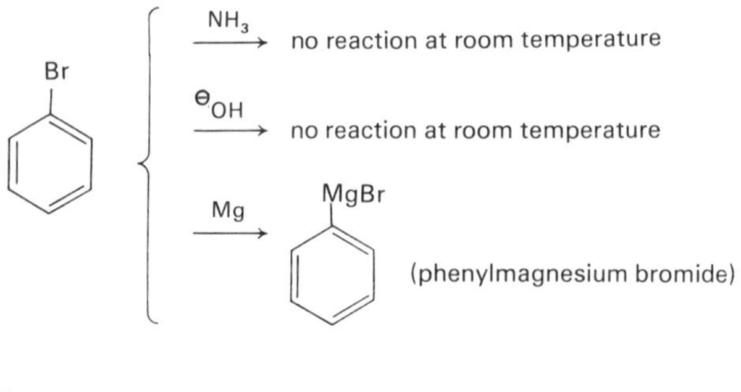

## Amino groups

Aniline contains a primary amino group. All amino groups have a lone pair of electrons which might be used to form bonds to protons. Aniline should therefore be basic, but it is found to be much less basic than methylamine. The proton transfer does not involve the trigonal carbon directly, but the presence of the benzene ring apparently influences the availability of the lone pair of electrons. One reason for this is that the picture drawn previously for the electron distribution in aniline is not completely adequate. The lone pair of electrons in aniline is actually involved in the delocalized system:

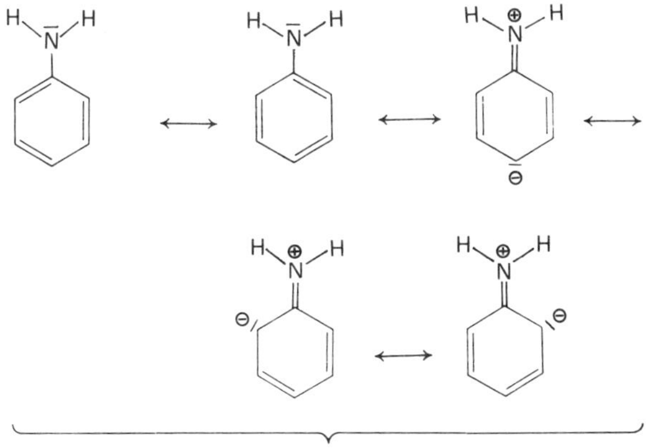

Aniline

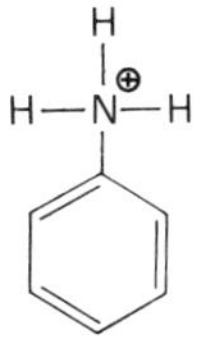

Anilinium ion
(phenylammonium
ion)

The true electron distribution is such that the electron density on the nitrogen is lower than the first drawing for aniline suggests, while that on the o- and p-positions of the ring is higher than the first drawing suggests.

The density of electrons around the nitrogen atom in aniline is less than that round the nitrogen atom in methylamine. The addition of a proton to the nitrogen atom in aniline is therefore less favourable than the protonation of methylamine.

We can look at the situation from another angle. The delocalization in the anilinium ion is less extensive than it is in aniline. Protonation of aniline causes a reduction in delocalization, whereas protonation of methylamine does not. Hence methylamine is the stronger base.

*Hydroxyl groups*

Phenol is a stronger acid than ethanol. In ethanol and in the ethoxide ion, the electrons are localized. Delocalization plays no part in determining their relative stabilities. In both phenol and the phenoxide ion the lone-pair electrons are not localized. The delocalization in phenol can be indicated by drawing electron distributions as was done on p. 130 opposite. A similar set can be drawn for the anion.

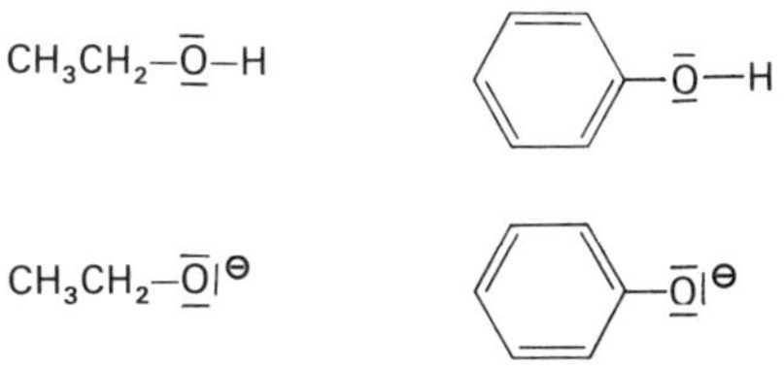

We can interpret the fact that the removal of a proton from phenol is easier than the removal of a proton from ethanol by assuming that the delocalization is more important and has a greater stabilizing effect in the phenoxide ion than in phenol.

*Alkyl groups*

Delocalization is possible for radicals as well as for substances with even numbers of electrons. When ethylbenzene is treated with bromine in the presence of light a reaction starts which results in the bromination of the ethylbenzene. The bromination of methane discussed in Chapter 8 proceeds by a similar pathway. Although there are five possible sites at which substitution could occur in ethylbenzene, only one monobrominated product is found:

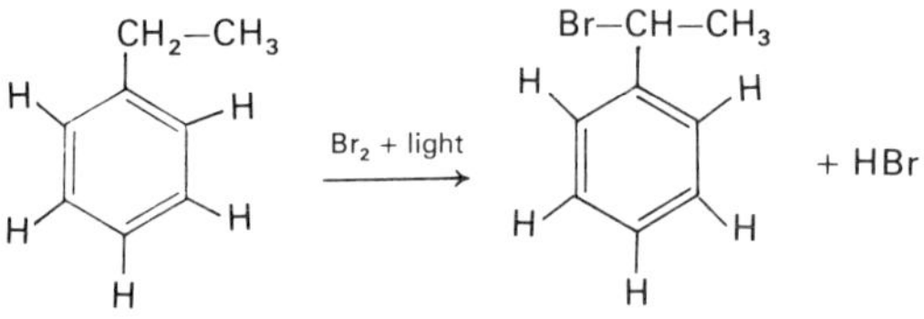

α-Bromoethylbenzene
(1-bromo-1-phenylethane)

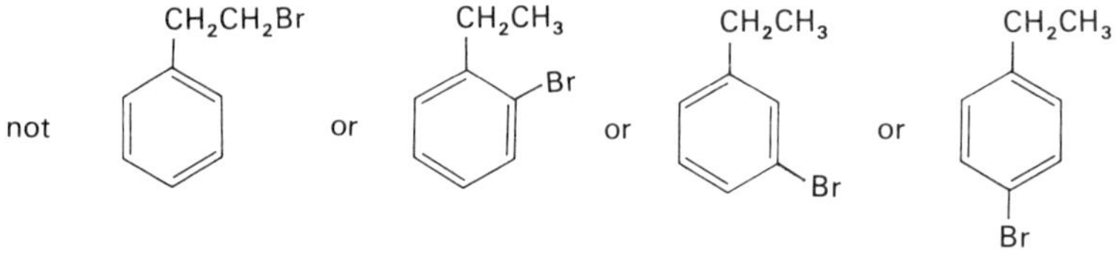

The reaction involves abstraction of a hydrogen atom from the hydrocarbon by a bromine atom. The radical produced then acquires a bromine atom from a bromine molecule and liberates another bromine atom:

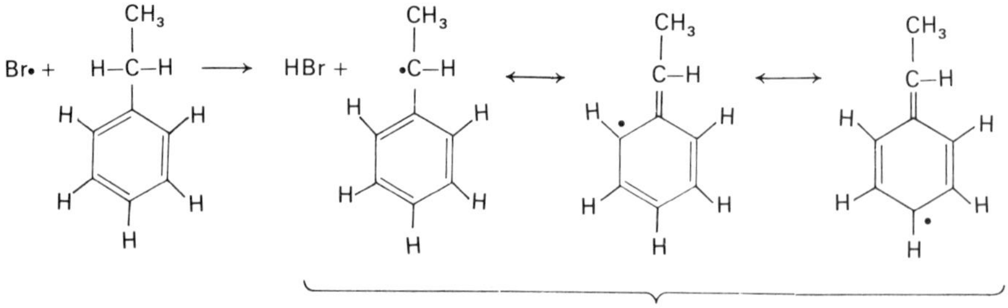

A delocalized radical

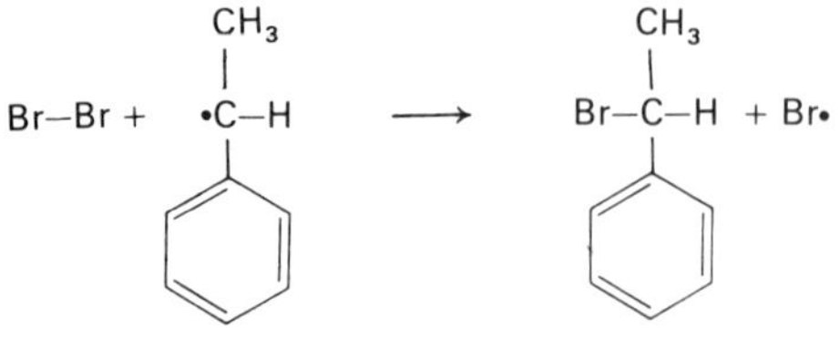

Abstraction of H · from the $CH_2$ group gives a radical which is delocalized. This radical is formed faster than its isomers since delocalization reduces the energy of the transition state leading to it.

*Problem 14–1. Draw pictures of the isomeric radicals which are not formed from the ethylbenzene. Are any of them delocalized? Would you expect the reaction of a bromine atom with ethylbenzene to be faster than its reaction with ethane?*

## 2. Reactions of the ring

Benzene might react with a nucleophile such as hydroxide ion to give an anion, or with an electrophile such as a proton to give a cation:

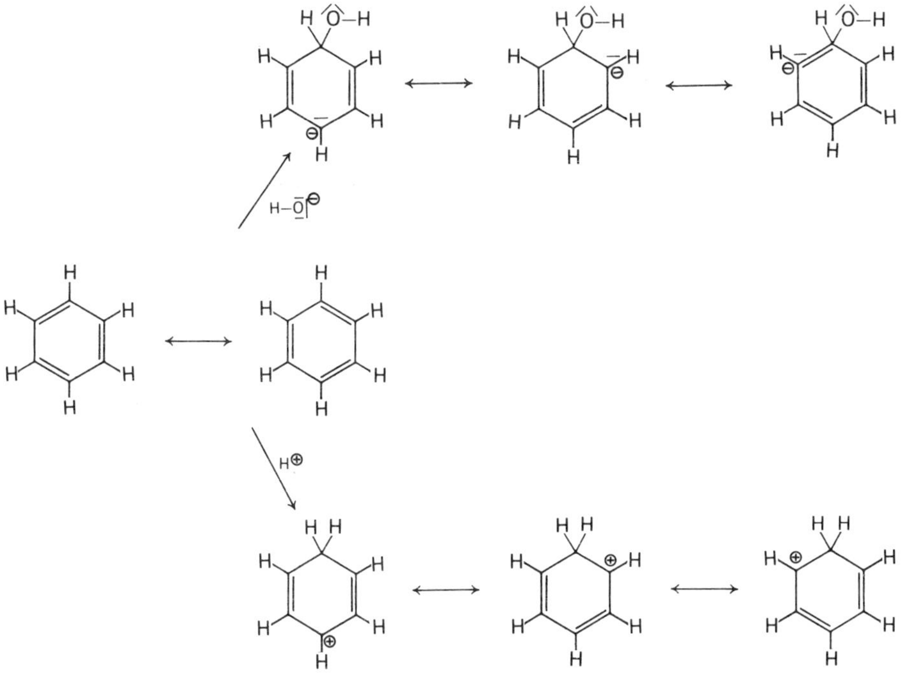

The anion and the cation both have one tetrahedral and five trigonal carbon atoms. They are both delocalized ions with the charge carried jointly by three of the five trigonal carbon atoms.

Benzene itself does not react with hydroxide ion, although benzenes containing substituents which stabilize the anion intermediate do react. Benzene does react with protons to give the cation above, but it is a very weak base and acids such as sulphuric acid in non-basic solvents (not water) are needed to donate a proton to benzene. We shall consider only reactions in which an electrophile adds to a benzene ring in the first step.

When electrophiles add to alkenes, the carbonium ion produced usually combines with a nucleophile resulting in overall addition of two groups to the double bond (see p. 119). If this happened to benzene, the product would be a cyclohexadiene, which cannot enjoy the same stabilization due to delocalization as benzene does. Loss of $X^{\oplus}$ from the carbonium ion might occur to give back the starting materials. The addition is reversible. However, an alternative fate for the carbonium ion is to lose a proton (see p. 108). Loss of a proton from the now tetrahedral carbon atom reproduces a benzene ring of six trigonal carbon atoms with no overall reduction in delocalization. The overall effect is a replacement of one hydrogen atom by the group X:

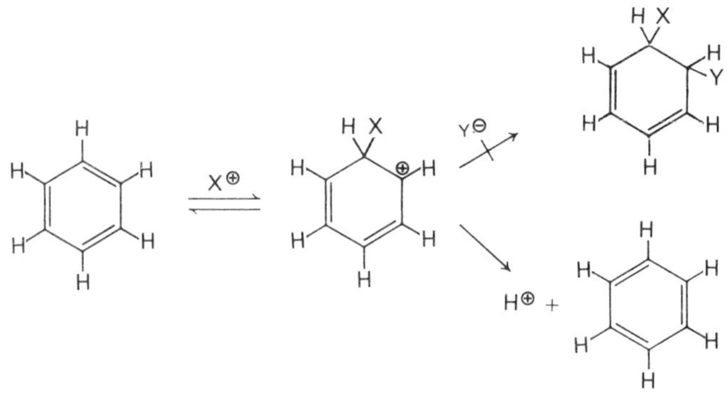

Overall addition of XY

Overall substitution of X for H

This last course leads to the observed products. The overall result is a substitution at carbon in which a trigonal carbon atom is attacked by an electrophile. It is sometimes called an electrophilic aromatic substitution. It is not the same as the $S_E$ reaction discussed in Chapter 10 in which bond formation and bond breaking occurred simultaneously at an initially tetrahedral carbon atom. This substitution at a trigonal carbon can be described as an addition–elimination process. Almost all substitutions at trigonal carbon atoms occur by addition–elimination. In this case it is an electrophile which adds and an electrophile which is eliminated.

Substitutions by addition–elimination in which a *nucleophile* attacks the trigonal carbon atom occur with some benzene derivatives and are common with carboxylic acid derivatives, as we shall see in Chapter 16.

In the diagram above, the reagent, the intermediate, and the product were each represented by one drawing only. We must not forget that all of these substances are delocalized:

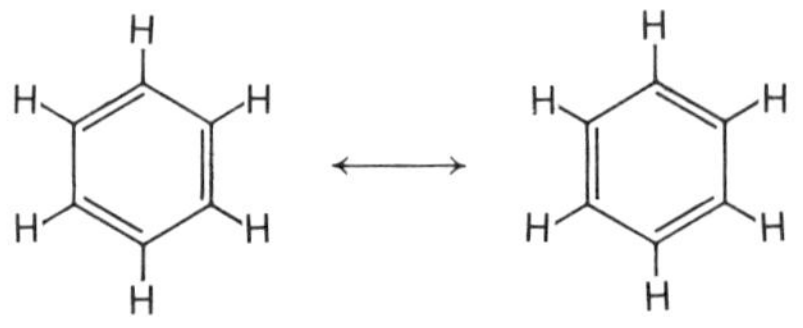

Reagent

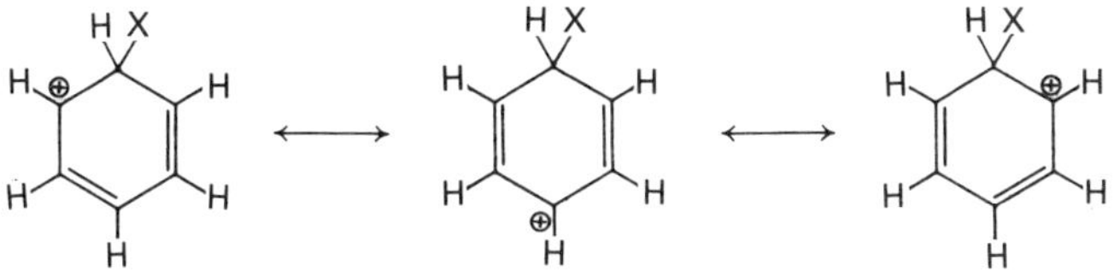

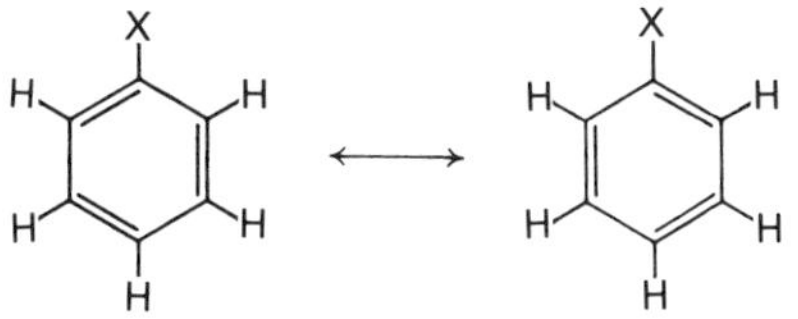

intermediate

product

In particular, the structure of the intermediate is such that the positive charge is shared among three of the trigonal atoms. If there are substituents on the ring, delocalization may be even more extensive and more drawings may be needed (see pp. 130, 131 and section 3 starting on p. 139).

*Problem 14–2. Draw out all the pictures which together represent the true electron distribution for the ion*

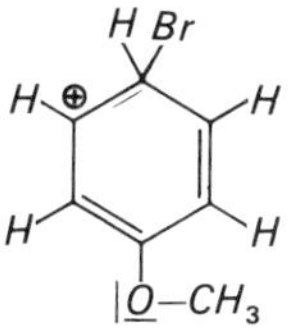

*formed during the bromination of anisole (methoxybenzene).*

We can now turn to the question of what kinds of electrophile, $X^\oplus$, can react with benzene or molecules containing a benzene ring. Metal ions such as $Na^\oplus$ do not react with benzene since they do not readily form a covalent bond to carbon, but $X^\oplus$ can be a proton or a carbonium ion, a nitronium ion, $NO_2{}^\oplus$, a diazonium ion, $R–\overset{\oplus}{N}\equiv N$, or an uncharged molecule such as $Br_2$ or $SO_3$. We have already seen that many of these electrophiles react with alkenes and the first step in the reaction of alkenes with electrophiles (see Chapter 13) is very like the first step in the reaction of benzenes with electrophiles.

*Introduction of halogen*
If benzene is treated with bromine, bromobenzene and HBr are formed very slowly:

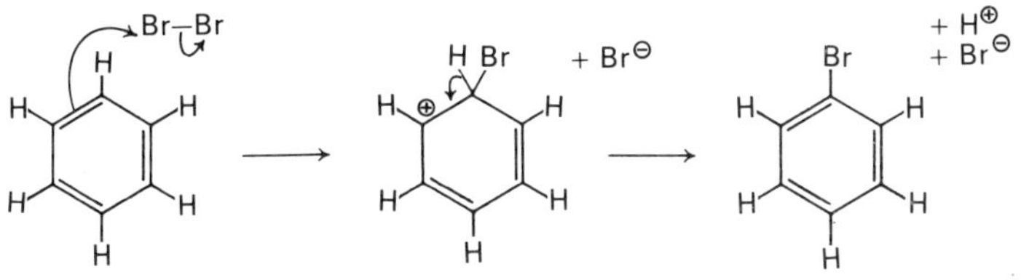

The reaction is much slower than the reaction of bromine with alkenes. The slow step is the first one, in which the benzenoid system of six trigonal carbons is destroyed. Some substituted benzenes such as phenol react rapidly with bromine.

*Problem 14–3. Look again at the discussion of phenol on p. 130 and suggest why the presence of the hydroxyl group makes the ring more nucleophilic. Which monobromophenol would you expect to be formed in the reaction of phenol with bromine?*

The reaction of benzene itself with bromine can be speeded up by converting the bromine into a better electrophile. This can be achieved by adding FeBr$_3$ or a similar Lewis acid (see pp. 64 and 88), which combines with the bromine to give a complex which can be written as

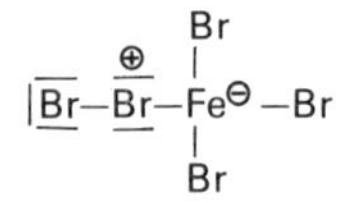

$$\overline{|Br}-\overset{\oplus}{Br}-\overset{\overset{\textstyle Br}{|}}{\underset{\underset{\textstyle Br}{|}}{Fe^{\ominus}}}-Br$$

Although the iron has a formal negative charge, the negative charge is largely carried by the bromine atoms attached to the iron since, being more electronegative, they draw electrons away from the iron. The benzene reacts with this complex faster than with bromine to give bromobenzene as before, plus the $^{\ominus}$FeBr$_4$ ion, which loses a bromide ion to re-form the FeBr$_3$ catalyst. The bromide ion combines with the proton displaced from the benzene ring and HBr gas escapes.

*Problem 14–4. Draw out the full mechanism of this reaction.*

*Introduction of alkyl and acyl groups*
Alkyl halides such as CH$_3$Br and acid chlorides (acyl chlorides) such as CH$_3$COCl are not sufficiently electrophilic to react with benzene. If FeBr$_3$, or better AlCl$_3$, is added, complexes are formed which can be written as

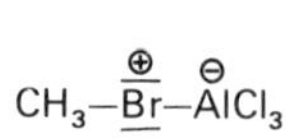

$$CH_3-\overset{\oplus}{Br}-\overset{\ominus}{AlCl_3} \qquad \text{and} \qquad CH_3-\overset{\overset{\textstyle O}{\|}}{C}-\overset{\oplus}{Cl}-\overset{\ominus}{AlCl_3}$$

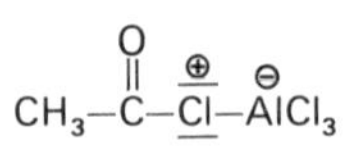

These complexes were mentioned on p. 88. Like the bromine complex above, they do react with benzene since they are much more electrophilic than CH$_3$Br:

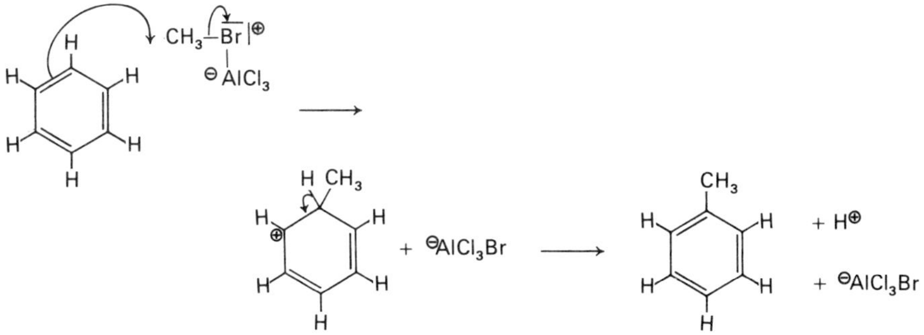

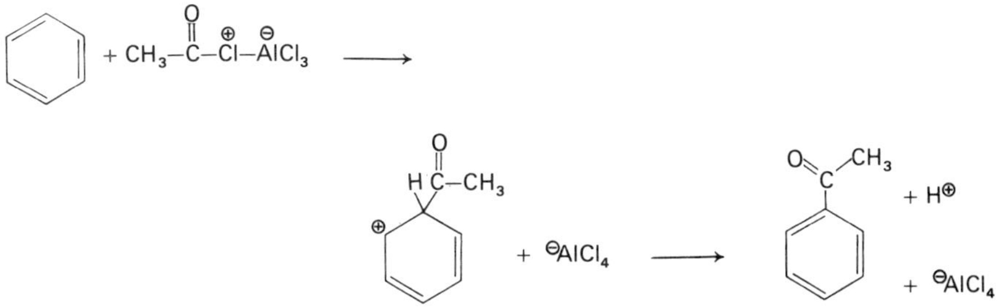

Acetophenone
(phenylethanone)

The displaced proton combines with a halide ion and escapes as HCl or HBr gas.
The $AlCl_3$ is thereby regenerated and is a catalyst for the reaction. In the second
case, most of the hydrogen atoms have not been drawn in explicitly. They are, of
course, still present. We shall use this shorthand drawing in the remainder of this
chapter. These reactions can be used to alkylate or acylate benzenes. An acyl

group has the general structure $R-\overset{\overset{\text{O}}{\|}}{C}-$. These reactions are called *Friedel–Crafts*
*reactions*, honouring the two chemists who discovered them.

*Introduction of nitro groups*
When nitric acid is dissolved in concentrated sulphuric acid it becomes protonated
and the oxonium ion decomposes to give water and the nitronium ion, $NO_2^{\oplus}$:

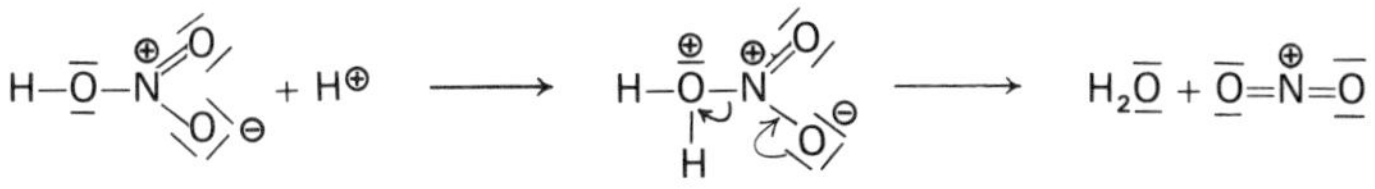

The nitronium ion reacts with benzenes to give nitrobenzenes. The process is
called *nitration*:

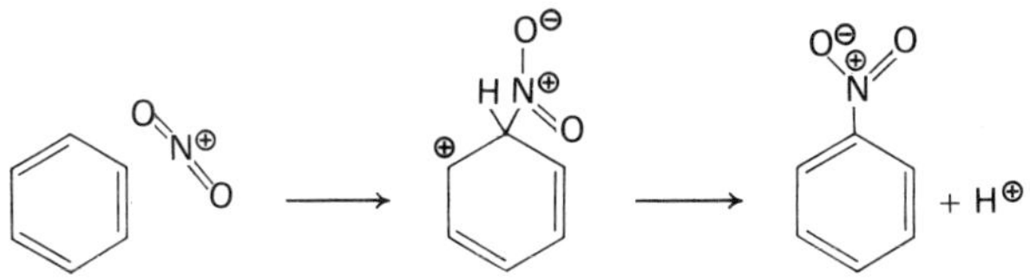

*Problem 14–5. Which common molecule is isoelectronic with the nitronium ion? Which of them is more electrophilic?*

*Introduction of sulphonic acid groups*

If benzene is treated with sulphuric acid or, better, with sulphuric acid containing extra sulphur trioxide, *sulphonation* of the benzene occurs. The product isolated is benzenesulphonic acid, $C_6H_5$–$SO_3H$, which is a strong acid:

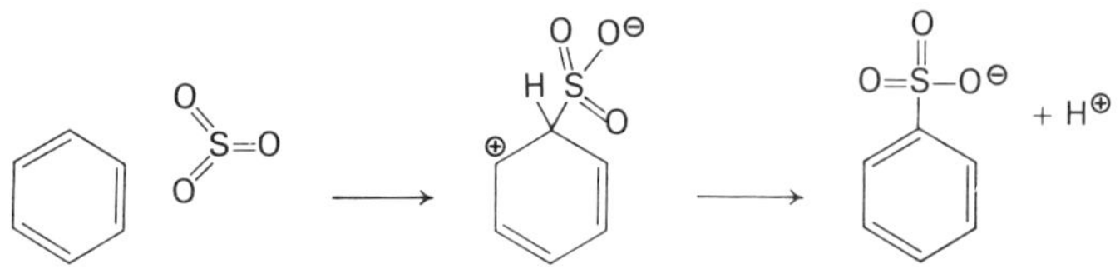

Benzenesulphonate ion

*Diazo coupling*

Diazonium ions such as benzenediazonium ion, $C_6H_5$–$\overset{\oplus}{N}\equiv N$, are weak electrophiles. The synthesis and some reactions of this ion were discussed on p. 108. Diazonium ions do not react with benzene but do react with certain substituted benzenes such as anilines and phenoxide ions. The outcome is a normal substitution.

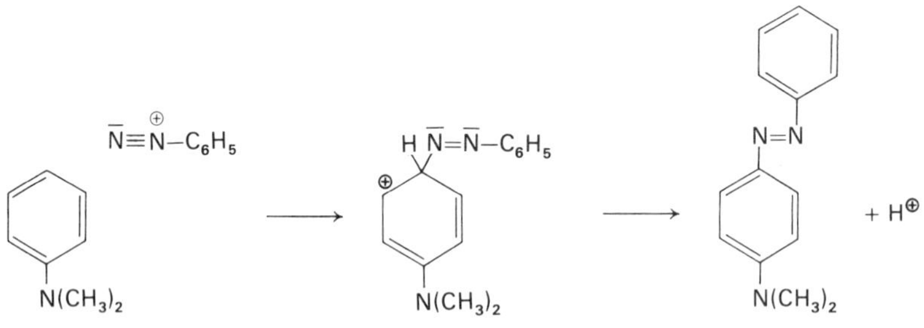

*NN*-Dimethylaniline

Note that in this reaction the diazonium ion does not lose nitrogen. The product is an azocompound, this particular one being 4-(dimethylamino)azobenzene. The *trans*-isomer is formed. Azobenzenes are usually coloured and can be used as dyes. This reaction by which they are made is called a *diazo coupling* reaction.

*Problem 14–6. Why does the nucleophilic benzene not become bonded to the nitrogen atom which is positive?*

The mechanism is the same in each case.

*Summary*

It is therefore possible to replace the hydrogen atoms of benzenoid compounds by several other groups.

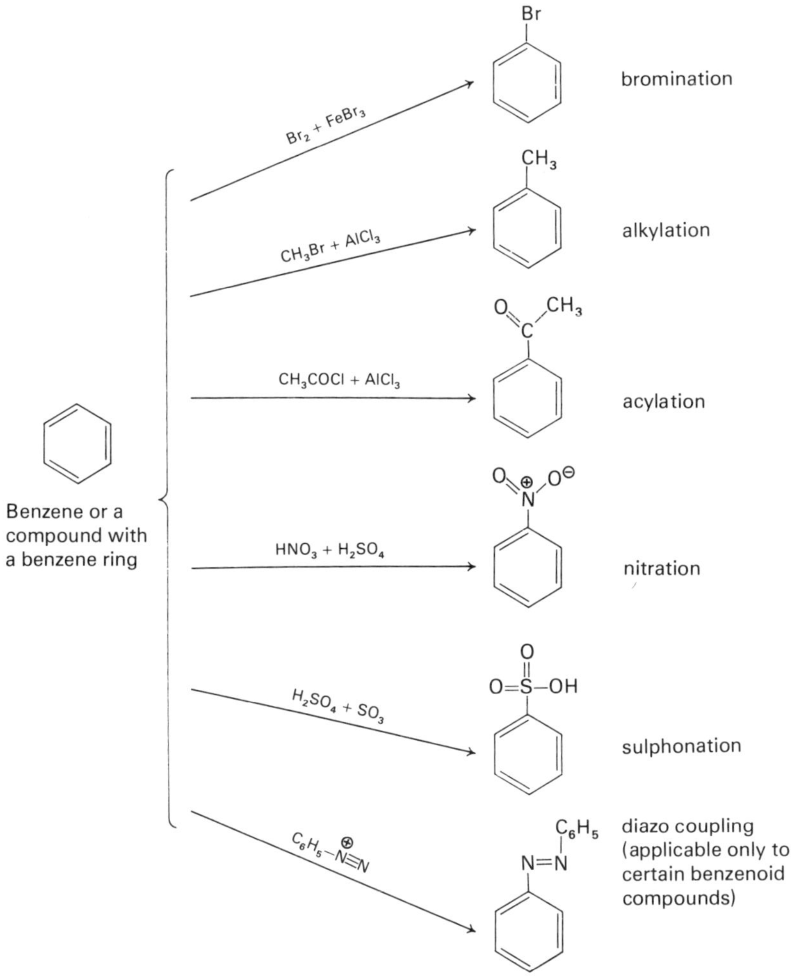

*Problem 14–7. Look back at the discussion of the hydration of propene (p. 120) and then suggest a method of making 2-phenylpropane from benzene and propene.*

### 3. Substitution at ring positions in derivatives of benzene

We have seen that a characteristic reaction of benzenoid compounds is substitution at a ring carbon atom by a mechanism which can be written in a general way as follows:

So far we have discussed the case where Y and Z are H, i.e. a hydrogen atom was the only group which could be displaced and there were no other substituents. If there are substituents in the ring initially, one of them rather than a hydrogen atom might be displaced. This is not common but is possible if the substituent lost (Y above) is fairly electropositive and forms a stable cation.

Even if the substituent is not displaced in preference to hydrogen, the situation is still complicated since the hydrogen atoms are now not all the same. Substitution of one of the hydrogen atoms in a benzene derivative:

could give the o-, the m-, or the p-disubstituted product or a mixture. We have already encountered two such cases—the bromination of phenol (P 14–3, p. 136) and the diazo coupling of dimethylaniline (p. 138).

The problem is very like that of the addition reactions of unsymmetrical alkenes. As we saw on p. 118, the structure of the final product is determined in the first step, in which the electrophile adds to the double bond of the alkene. The site of addition is determined by the substituents attached to the double bond of the alkene. If the reaction is under thermodynamic control, the most stable cation will be formed. If the reaction is under kinetic control, the cation formed is the one formed fastest, i.e. the one formed via the transition state of lowest energy. The substituents on the alkene can stabilize the product cations, and the transition states leading to them, either because they permit delocalization or because their electronegativity is low.

The situation with substituted benzenes is very similar. The first step in the substitution process is addition of an electrophile to the ring to give a cation. Several cations may be possible. It is found that most substitutions of benzenes are under kinetic control so the cations formed are those which are formed the fastest, i.e. those formed via the transition states of lowest energy.

The transition states differ only in the position of the *original* substituents relative to the site of attack. The energies of the transition states depend on the stabilizing or destabilizing effect of the original substituents. In fact, the distribution of charge and the possibilities of delocalization in the transition states are very like those in the cations produced. The effect of substituents on the site of substitution can therefore be predicted by looking at the cations since they are similar to the transition states.

If the substituents in any one cation can stabilize it and stabilize the transition state leading to it by, for example, allowing more extensive delocalization, then that cation will be formed faster than its isomers and the product derived from it will be a major product.

If the substituents in any one cation destabilize it and therefore destabilize the transition state leading to it by, for example, withdrawing electrons from one of the ring positions which is acquiring a positive charge, then that cation will be formed more slowly than its isomers and the product derived from it will be a minor product.

We shall consider only one case here, the bromination of anisole (methoxybenzene). There are four different sites at which the bromine molecule might attack and four different intermediates are possible (see p. 142).

The cation A might be formed but it cannot readily break down to give $CH_3-O^{\oplus}$. Any A formed reverts to starting material. Ions B, C, and D can all break down with expulsion of $H^{\oplus}$ and give bromoanisoles. The relative rates of formation of B, C, and D determine the proportions of the isomeric bromoanisoles formed. Intermediates C and D are more delocalized than B (or A). One more picture can be drawn for them. The ability of the oxygen atom in C and D but not B to carry part of the positive charge is significant even in the transition state leading to these ions. C and D are formed faster than B. The observed product is a mixture of 2-bromoanisole and 4-bromoanisole plus a trace amount of 3-bromoanisole.

The $CH_3O$ substituent in this case accelerates the formation of two of the possible products relative to the others (and also relative to the bromination of benzene). Bromine reacts rapidly at the 2- and 4-positions of anisole and only very slowly at the 3-position of anisole or at any position in benzene.

The sites of attack by electrophiles on other substituted benzenes can be predicted by the effects the substituents will have on the relative stability of the transition states which are possible in the first step.

The following table summarizes the observed dominant directing effect of some substituents. These observations are in general accord with the explanation given above.

| Substituent initially present | Position(s) attacked (relative to the initial substituent) |
| --- | --- |
| $-NO_2, -SO_3H, -\overset{\displaystyle O}{\overset{\displaystyle \|}{C}}-R, -\overset{\displaystyle O}{\overset{\displaystyle \|}{C}}-OR$ | *meta* |
| $-Cl, -Br, -OR, -NR_2, -R$ | *ortho* and *para* (usually a mixture of both) |

The group R in the table can be H, alkyl, or aryl. The nature of the electrophilic reagent is largely immaterial.

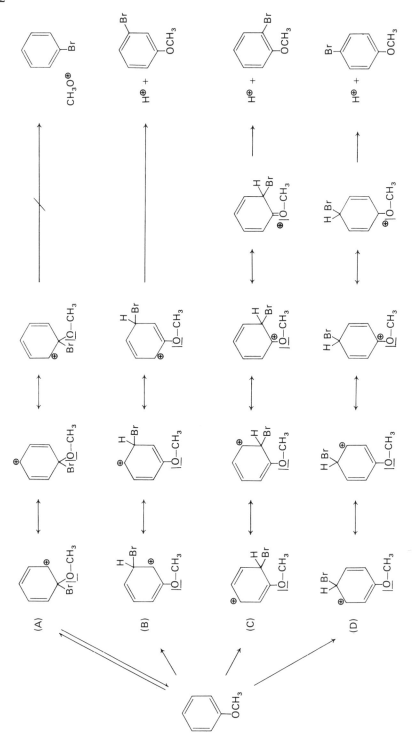

*Problem 14–8. Why does the benzenediazonium ion react with aniline but not with benzene? Is your explanation in accord with the observation that the product is the result of addition of the electrophile to the position* para *to the amino group?*

*Problem 14–9. It was suggested earlier that groups other than hydrogen could be displaced in substitution by electrophiles at benzenoid carbon atoms. Suggest a mechanism for the reaction below:*

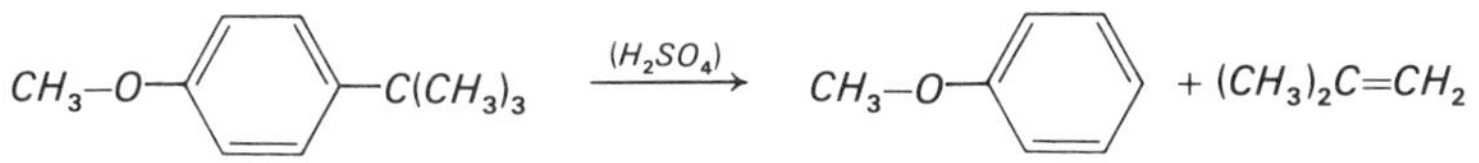

$$CH_3-O-\!\!\!\bigcirc\!\!\!-C(CH_3)_3 \xrightarrow{(H_2SO_4)} CH_3-O-\!\!\!\bigcirc + (CH_3)_2C=CH_2$$

*Anisole*

*How will the protons react with this ether? To what atoms might the protons add? What delocalized cationic adducts can you draw? What is the leaving group in the substitution? What happens to it?*

*Problem 14–10. Suggest a method of making 2,4,6-trideuterioanisole from anisole. Deuterium is an isotope of hydrogen. It is available as deuterium chloride gas, $^2HCl$, and deuterium oxide, $^2H_2O$.*

### ANSWERS TO PROBLEMS IN CHAPTER 14

*Problem 14–1.*

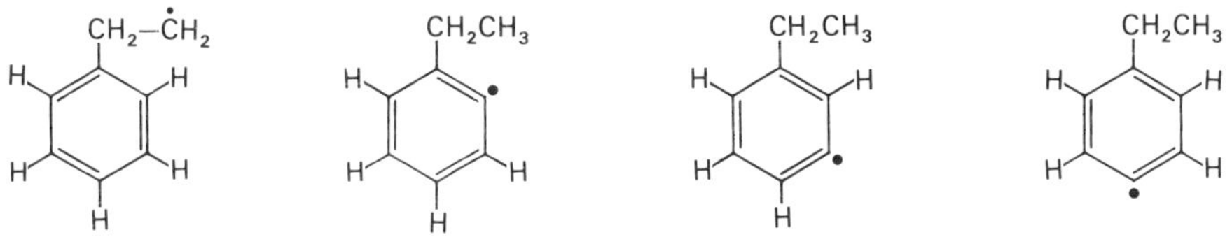

None of the above is delocalized. The bromination of ethylbenzene should be faster than that of ethane since the presence of the phenyl group accelerates the removal of H· in the first step. The delocalization present in the radical $C_6H_5\dot{C}HCH_3$ also influences the energy of the transition state leading to that radical.

*Problem 14–2.* See pp. 141 and 142.

*Problem 14–3.* The full picture of phenol discussed on p. 130 shows that the *o*- and *p*-positions have higher electron density than the carbon atoms of benzene and so might be expected to react with bromine faster than the *m*-positions and faster than benzene. This should lead to attack selectively at the *o*- and *p*-

positions. The observed products are o- and p-bromophenol. A more detailed analysis of this type of situation is given on p. 141.

*Problem 14–4.*

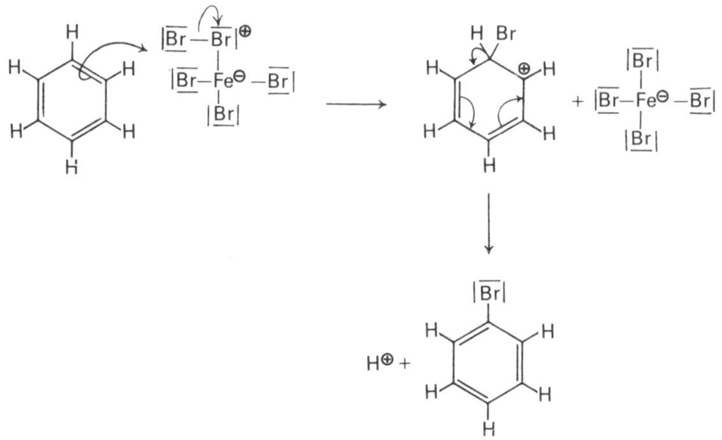

*Problem 14–5.* $\overline{O}=C=\overline{O}$ is isoelectronic with $\overline{O}=\overset{\oplus}{N}=\overline{O}$. Because of the charge, the nitronium ion is much more electrophilic than carbon dioxide.

*Problem 14–6.* The positive nitrogen cannot accept electrons to form another bond since that would require it to accommodate a total of 10 electrons around it. It can acquire one of its bonding pairs to be a lone pair if the nucleophile attacks the neighbouring nitrogen atom.

*Problem 14–7.* Friedel–Crafts alkylation of benzene by $CH_3-\overset{\oplus}{C}H-CH_3$ or its equivalent will give 2-phenylpropane. This cation itself could be made from propene plus a proton (see p. 118). The proton source would need to introduce no nucleophiles which would compete with the benzene. Therefore, mix benzene, propene and a trace amount of an acid such as $H_2SO_4$.

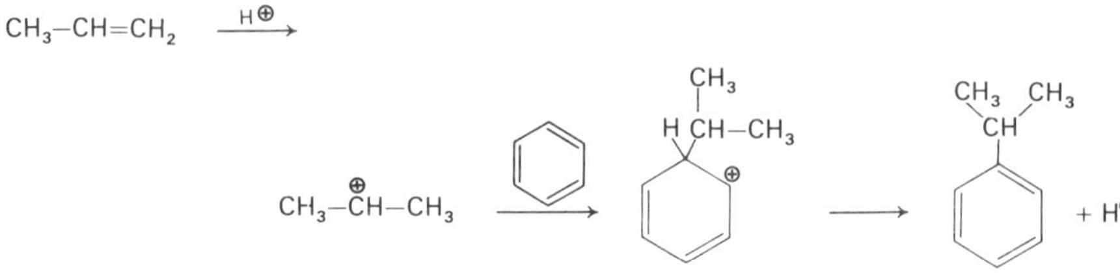

$$CH_3-CH=CH_2 \xrightarrow{\ H^\oplus\ }$$

Overall:

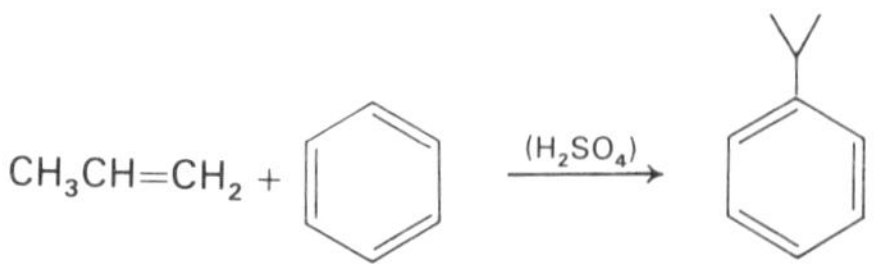

Alternatively:

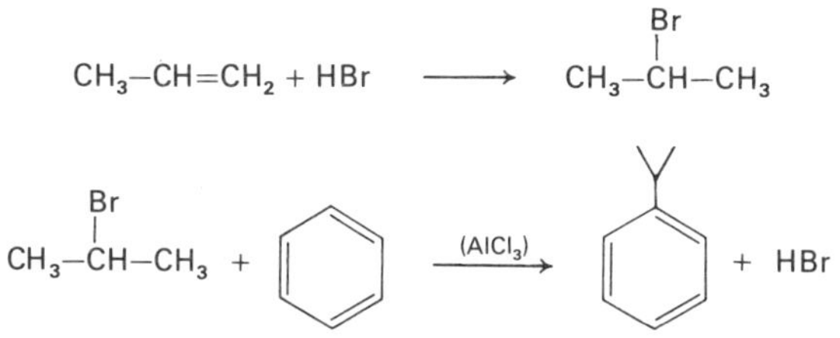

$$CH_3-CH=CH_2 + HBr \longrightarrow CH_3-\overset{Br}{\underset{|}{CH}}-CH_3$$

$$CH_3-\overset{Br}{\underset{|}{CH}}-CH_3 + \text{⬡} \xrightarrow{(AlCl_3)} \text{⬡} + HBr$$

*Problem 14–8.* The amino group accelerates attack by electrophiles but only at the *o-* and *p*-positions. The intermediates

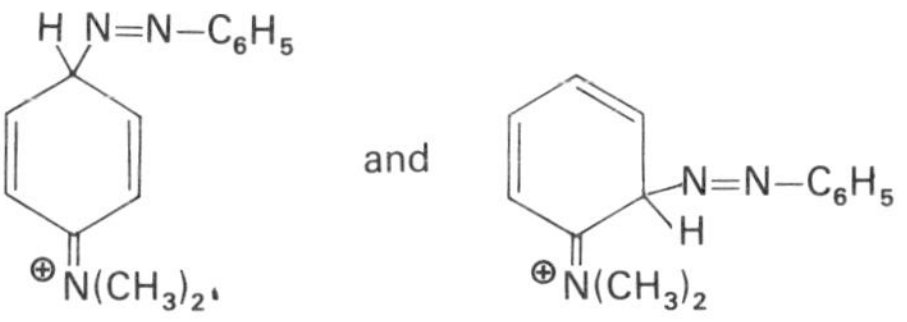

and

are more delocalized than

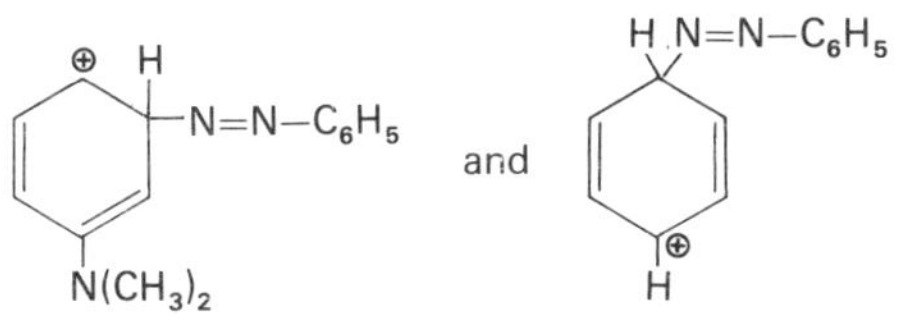

and

This delocalization is important in the transition states leading to these ions so the first two are formed faster. In this case the intermediate resulting from *p*-attack is formed fastest of all.

*Problem 14–9.* Protonation could give cations A, B, C, or D. B and D are most extensively delocalized and are formed fastest.

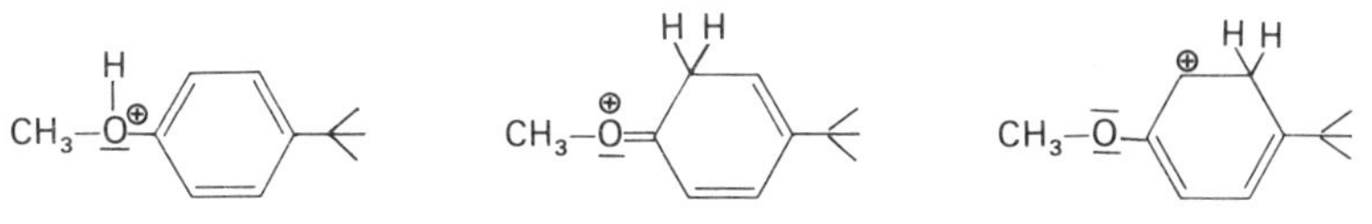

A      B      C

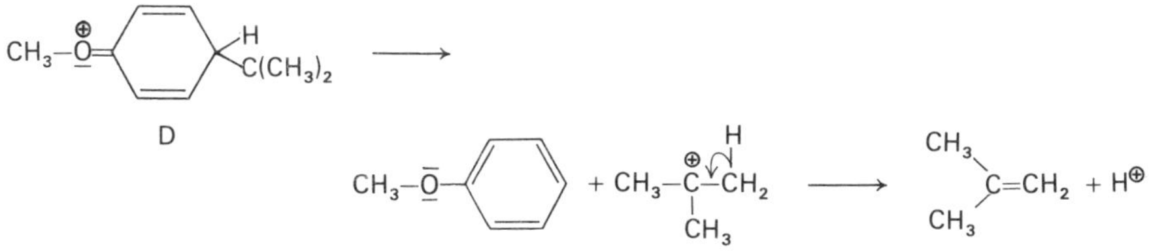

D

$$CH_3-\bar{\overset{\oplus}{O}}-\text{⬡} + CH_3-\overset{H}{\underset{|}{\overset{\oplus}{C}}}{}^{\curvearrowleft}CH_2 \longrightarrow \overset{CH_3}{\underset{CH_3}{>}}C=CH_2 + H^{\oplus}$$

146

In the absence of a nucleophile which might add to them, A, B, and C can only revert to starting materials. However, loss of the $(CH_3)_3C^\oplus$ cation from D is possible and indeed fast. This cation is stable (see p. 117), but will eventually lose a proton to give an alkene (since there are no nucleophiles available except the anisole).

The reaction is, in fact, reversible. An equilibrium will be established. If the alkene is allowed to boil out of the reaction vessel (it is a gas, p. 113), the equilibrium concentrations cannot be established and anisole will be formed in 100% yield. If a high concentration of alkene were maintained by constantly passing it through the solution, the anisole could be alkylated (see Problem 14–7).

*Problem 14–10.* The group $CH_3$–O– accelerates attack by nucleophiles such as $^2H^\oplus$ at *o*- and *p*-positions (see bromination of anisole). Therefore, treat the anisole with excess of a source of $^2H^\oplus$ such as $^2HCl$. Substitution of $^2H$ for $^1H$ will occur at the 2-, 4-, and 6-positions. The introduction of the $^2H$ will not affect the positions attacked subsequently. Eventually, substitution at all three sites will be achieved, giving 2,4,6-trideuterioanisole.

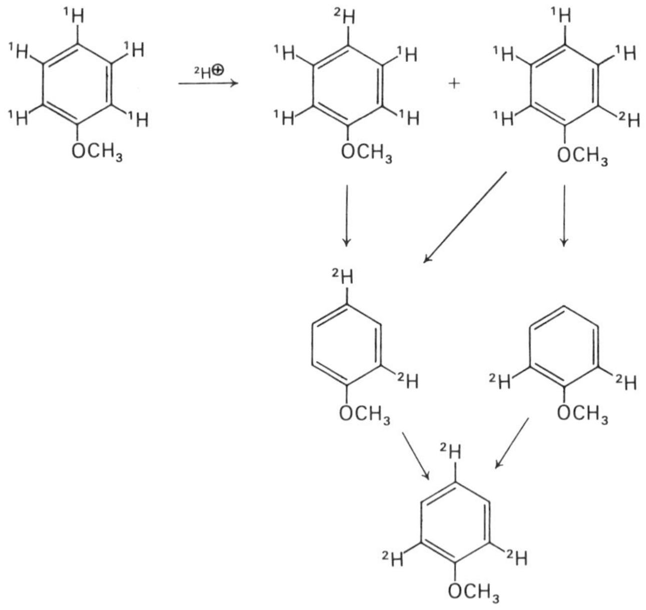

# Chapter 15

## Carbonyl Compounds

Compounds containing the carbonyl group, $>$C=O, attached to hydrogen or carbon only, are either aldehydes or ketones. The ketones may be symmetrical or unsymmetrical.

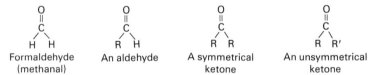

| Formaldehyde (methanal) | An aldehyde | A symmetrical ketone | An unsymmetrical ketone |

R and R′ in the above structures can be any alkyl or aryl group.

Aldehydes are often named after the related carboxylic acid but can also be named systematically using the suffix -al. Ketones can be named systematically using the suffix -one and in other ways.

$CH_3CH_2CH_2CHO$    $CH_3CH_2COCH_3$    $CH_3COC_6H_5$

Butanal (butyraldehyde)

Butan-2-one (ethyl methyl ketone)

1-Phenylethanone (acetophenone; methyl phenyl ketone)

Note that when the carbonyl compounds are written in this contracted in-line manner, the carbonyl group is always written with the oxygen immediately after (to the right of) its carbon atom, except for aldehydes, which are written as HCOR or RCHO. Methanal (formaldehyde) can be written as HCHO or $CH_2O$.

The four atoms $\overset{X}{\underset{Y}{>}}$C=O in any carbonyl compound lie in one plane and the X–C–O and Y–C–O angles are very close to 120° (see p. 35). The carbonyl group is fairly polar owing to the electronegativity difference between carbon and oxygen, and the oxygen can take part in hydrogen bonding.

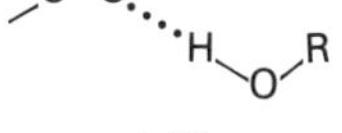

$$>\!C\!=\!O\cdots H\!-\!O\!-\!R$$

Carbonyl compounds with small alkyl groups are rather volatile liquids and are soluble in water. The smallest, methanal (formaldehyde) ($CH_2O$), is a gas with a choking smell. Propan-2-one (acetone) is a useful solvent which can dissolve many organic substances and can also dissolve water. Ketones and aldehydes with larger alkyl or aryl groups are less volatile and less soluble in water.

| Formula | Systematic name | Trivial name | B.p. (°C) | Solubility in water |
|---|---|---|---|---|
| $CH_2O$ | Methanal | Formaldehyde | −21 | Completely soluble |
| $CH_3CHO$ | Ethanal | Acetaldehyde | 21 | Completely soluble |
| $CH_3COCH_3$ | Propan-2-one | Acetone | 56 | Completely soluble |
| $CH_3CH_2COCH_3$ | Butan-2-one | Methyl ethyl ketone | 80 | Slightly soluble |
| $C_6H_{10}O$ | Cyclohexanone | | 157 | Insoluble |
| $C_6H_5CHO$ | | Benzaldehyde | 179 | Insoluble |
| $C_6H_5COCH_3$ | Phenylethanone | Acetophenone | 204 | Insoluble |

Aldehydes are rather easily oxidized by oxygen, $Ag^+$, $Cu^{2+}$, and more vigorous reagents to give carboxylic acids. In other words, aldehydes are reducing agents. The reduction by aldehydes of silver ion to silver, or of copper(II) to copper(I), provides a convenient distinguishing test since most ketones cannot be oxidized by these reagents. The equation below shows what happens when an aldehyde reacts with Fehling's solution or Benedict's solution, both strongly alkaline solutions containing a compound of Cu(II).

$$RCHO + 2Cu^{2+} + 4OH^{\ominus} \longrightarrow RCOOH + Cu_2O\downarrow + 2H_2O$$

(as its salt)   red precipitate

## ADDITION OF HYDROGEN

Carbonyl compounds, like alkenes, are unsaturated and addition reactions are to be expected. For example, carbonyl compounds can be hydrogenated catalytically to give alcohols:

$$\text{>C=O} + H_2 \xrightarrow{(Pt)} \text{>C}\substack{H\\OH}$$

The carbonyl group usually takes up hydrogen more slowly than does a C=C double bond.

## SITE OF ADDITION OF ELECTROPHILES OR NUCLEOPHILES

The ionic addition reactions of carbonyl compounds can be compared with those of alkenes. The ends of an alkene double bond differ only if the substituents are different and the site of addition of electrophiles and nucleophiles is determined by the nature of the substituents. In aldehydes and ketones there is a very distinct difference between the ends of the double bond and the presence of the oxygen atom controls the site of addition.

It is found that both electrophiles and nucleophiles can add to carbonyl groups. Electrophiles always add to the oxygen and nucleophiles always add to the carbon.

The addition of an electrophile such as a proton to a carbonyl group is a reversible process which is under thermodynamic control. The more stable product predominates. Addition of the proton to carbon would give a cation with a positive charge localized on an oxygen atom which has not got a complete valency shell of electrons:

$$\underset{/}{\overset{\backslash}{C}}\underset{\overset{|}{H}}{-}\bar{O}^{\oplus}$$

Addition of the proton to oxygen, on the other hand, produces a delocalized ion in which the charge is spread between carbon and oxygen:

$$\underset{/}{\overset{\backslash}{C}}{=}\underline{\overset{H^{\oplus}}{O}} \quad \longrightarrow \quad \underset{/}{\overset{\backslash}{C}}{=}\overset{\oplus}{O}\underset{H}{\overset{}{\diagdown}} \quad \longleftrightarrow \quad \underset{/}{\overset{\oplus\backslash}{C}}{-}\bar{O}\underset{H}{\overset{}{\diagdown}}$$

This delocalized product is always the one formed.

The addition of most nucleophiles to carbonyl groups is under kinetic control. The product formed is that which is formed fastest, i.e. that formed via the transition state of lowest energy.

$$\underset{/}{\overset{\backslash}{C}}{=}\underline{O} \quad \overset{H^{\ominus}}{\rightarrow} \quad \underset{/}{\overset{\ominus\backslash}{C}}{-}\bar{O}{-}H$$

$$\underset{/}{\overset{\backslash}{C}}{=}\underline{O} \quad \overset{\ominus H}{\rightarrow} \quad \underset{/}{\overset{\backslash}{\underset{\overset{|}{H}}{C}}}{-}\bar{O}|^{\ominus}$$

In the first case negative charge is building up on the carbonyl atom as the reaction proceeds. In the second case the more electronegative oxygen atom is becoming negatively charged. The second course is therefore the easier and the faster.

Addition of an electrophile to a carbonyl compound therefore gives a carbonium ion. This may combine with a nucleophile, the overall effect being addition of two groups to the double bond. Alternatively, the nucleophile may add first to give an alkoxide ion, which then combines with an electrophile. In either case the sites of attachment of the electrophilic group (X) and the nucleophilic group (Y) are the same.

$$\underset{/}{\overset{\backslash}{C}}{=}O \quad
\begin{matrix}
\overset{X^{\oplus}}{\nearrow} & \underset{/}{\overset{\oplus\backslash}{C}}{-}O{-}X & \overset{Y^{\ominus}}{\searrow} \\
& & \\
\underset{Y^{\ominus}}{\searrow} & \underset{\overset{|}{Y}}{\overset{\backslash}{C}}{-}O^{\ominus} & \overset{X^{\oplus}}{\nearrow}
\end{matrix}
\quad \underset{\underset{Y}{\overset{|}{\phantom{x}}}}{\overset{\overset{|}{\phantom{x}}}{\overset{\backslash}{C}}}\underset{X}{\overset{\text{—}O}{}}
\qquad
\begin{matrix}
\text{Overall addition of} \\
\text{YX to C}{=}\text{O}
\end{matrix}$$

## ADDITIONS IN WHICH THE NUCLEOPHILE ADDS FIRST

### Carbanions

Carbonyl compounds can react with good nucleophiles in a process initiated by addition of the nucleophile. The addition always occurs at the carbon end and produces an alkoxide ion. Usually the reaction is completed by adding water or some better proton source which converts the alkoxide ion into an alcohol. Thus, acetone reacts with sodium cyanide followed by ammonium chloride to give a cyano alcohol called a *cyanohydrin*, which is an adduct of acetone + HCN:

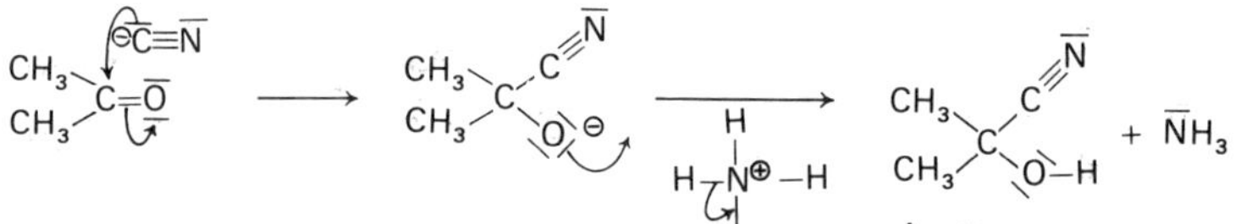

Acetone cyanohydrin

Cyanohydrins can be made from most ketones and aldehydes.

Cyanide ion is, of course, a carbanion and other carbanions behave similarly. The ion $^{\ominus}CH_2NO_2$ can be made from nitromethane, $CH_3NO_2$, and potassium hydroxide. The ion is very stable owing to delocalization and to the formal positive charge on the nitrogen which draws electrons away from the carbon atom.

The salt $K^{\oplus}\ ^{\ominus}CH_2NO_2$ is completely ionic. This carbanion adds to acetone to give an alkoxide ion. Addition of acid then produces a nitro alcohol:

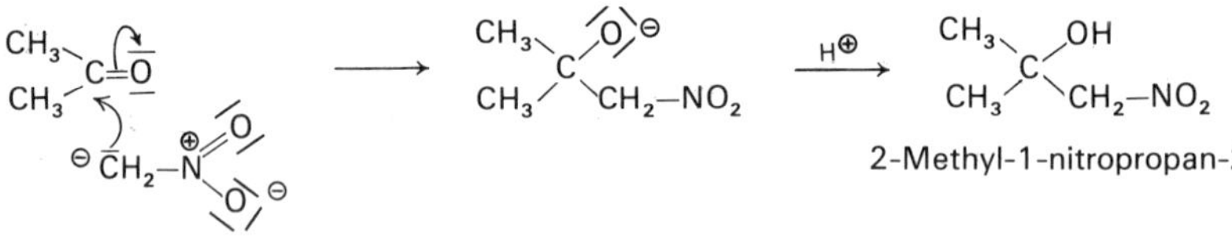

2-Methyl-1-nitropropan-2-ol

Covalent organometallic compounds also react with carbonyl groups. Aldehydes and ketones react with Grignard reagents and alkyllithiums. The metal alkoxides formed first can be converted into alcohols by adding acid.

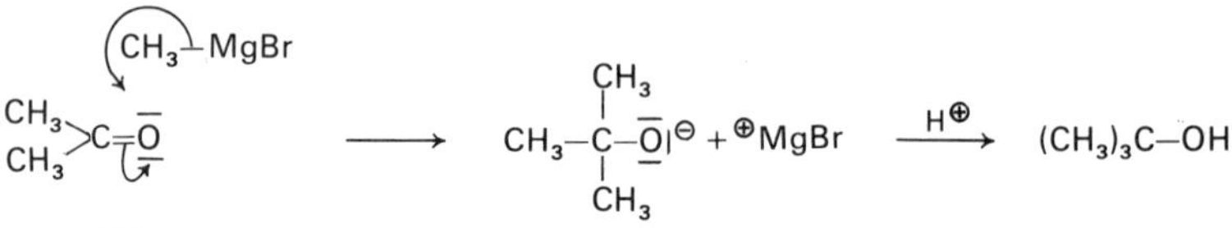

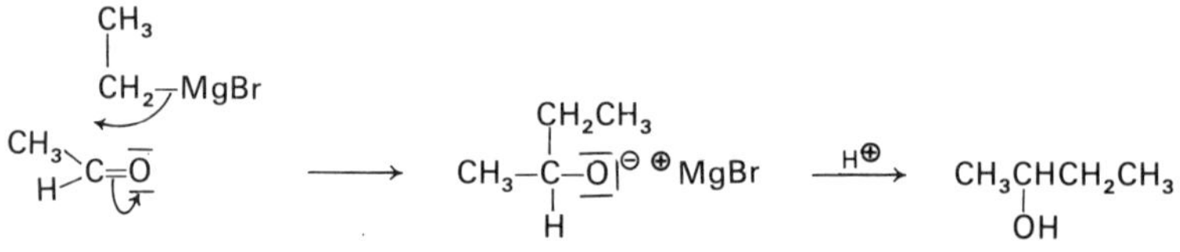

*Problem 15–1. How would you convert benzyl bromide ($C_6H_5CH_2Br$) into 2-phenylethanol?*

## Sulphite ion

Acetaldehyde combines with sodium hydrogen sulphite to give the sodium salt of 1-hydroxyethanesulphonic acid. Sulphonic acids have the general structure $R–SO_3H$ (see benzenesulphonic acid, p. 127).

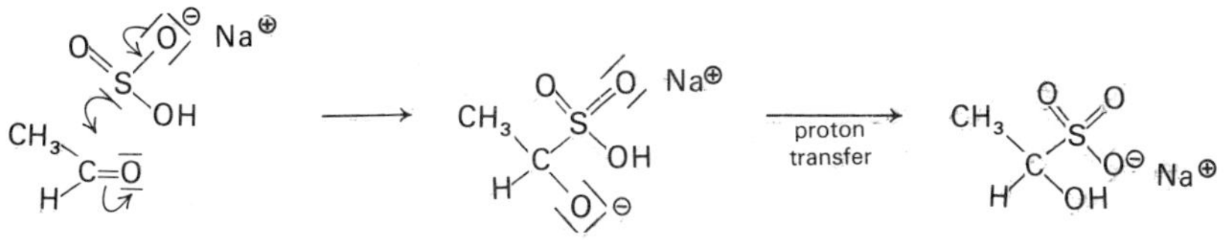

The salt is an adduct of the aldehyde and $NaHSO_3$. It is sometimes called the bisulphite adduct of the aldehyde. Check carefully the valency electrons of the sulphur, which has a lone pair to begin with and is acting as a nucleophile. Most aldehydes and some ketones will react with $NaHSO_3$ in this way. The adducts, being salts, are soluble in water.

*Problem 15–2. Cyclohexanol and cyclohexanone are both liquids, insoluble in water, and with similar boiling points. Suggest a method of removing trace amounts of cyclohexanone from cyclohexanol.*

## Hydride ion

Carbonyl compounds can be reduced to alcohols by catalytic hydrogenation but a more convenient and more selective way to do this is by treatment with sources of hydride ion such as the tetrahydroborate, $^{\ominus}BH_4$, or tetrahydroaluminate, $^{\ominus}AlH_4$, ions. Treatment of carbonyl compounds with sodium borohydride, $NaBH_4$, in water or with aluminium lithium hydride (lithium tetrahydroaluminate), $LiAlH_4$, in ether followed by addition of acid reduces them to alcohols under very mild conditions:

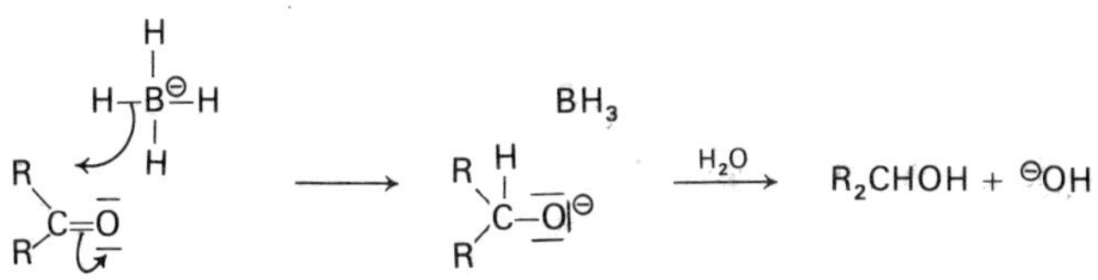

In fact, all four of the BH bonds are usable. The overall equation can be written as

$$4R_2CO + NaBH_4 + 3H_2O \rightarrow 4R_2CHOH + NaOB(OH)_2$$

Sodium dihydrogen borate

The tetrahydroaluminate ion behaves in exactly the same way. It is easy then to convert carbonyl compounds into alcohols or alcohols into carbonyl compounds (p. 100).

*Problem 15–3. How would you convert pent-4-enal to pentanal and to pent-4-en-1-ol?*

## ADDITIONS IN WHICH THE ELECTROPHILE ADDS FIRST

All of the nucleophiles discussed above which add to ketones are strong nucleophiles. Weaker nucleophiles such as alcohols can be made to add to carbonyl groups but the alcohol does not add to the carbonyl group as such in the first step. It is necessary first to make the carbonyl group more electrophilic.

In the chapter on benzene derivatives, it was shown that the relatively weak nucleophile, benzene, will react with weak electrophiles such as bromine or methyl halides if these are first treated with a second electrophile such as $FeBr_3$. In the same way, electrophiles such as metal ions or $BF_3$ can be used to improve the electrophilicity of the carbonyl carbon of carbonyl compounds but the most convenient one to use is $H^\oplus$.

As discussed above, protons add to carbonyl groups to give a delocalized ion which will be represented here by one picture:

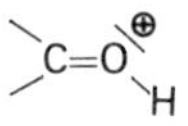

$$\text{\large$\Big\rangle$}C{=}\overset{\oplus}{O}\diagdown_H$$

This ion is much more electrophilic than the original ketone and will react with weak nucleophiles such as amines, alcohols, and water, forming a new bond to the carbon (see p. 97).

Protonated ketones do not react with halide ions since the equilibrium constant for

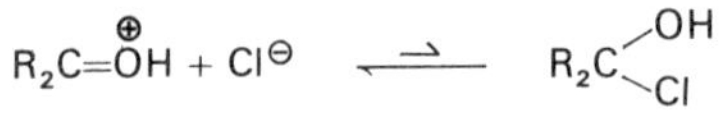

$$R_2C{=}\overset{\oplus}{O}H + Cl^\ominus \;\rightleftharpoons\; R_2C\diagup^{OH}_{Cl}$$

is very small. Therefore HCl, HBr, etc., cannot be added to carbonyl groups. In the cases discussed below, the equilibrium constants are larger.

### $H^\oplus$ plus alcohols

When cyclohexanone is treated with methanol and a small amount of acid, the first product is a 1:1 adduct which is called a hemiacetal:

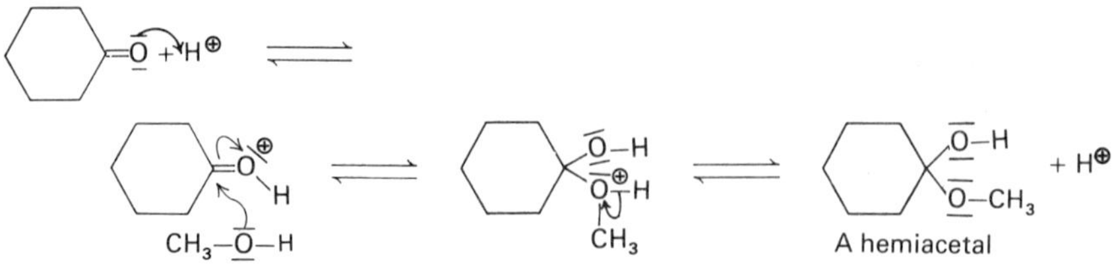

The overall reaction is addition of methanol to the ketone. The acid is a catalyst. The reaction involves addition of the electrophile, $H^\oplus$, first, then addition of the nucleophile. This whole process is reversible and the equilibrium lies to the right only if the concentration of methanol is high. If the hemiacetal is dissolved in water containing protons, the reaction reverses and the ketone is re-formed.

The hemiacetal is in fact not usually isolated from this reaction. If there is plenty of methanol, the observed product is cyclohexanone dimethyl acetal. This requires a second mole of methanol and involves formation of a second ether group. The mechanism involves an $S_N$ displacement but bond formation and bond breakage are not simultaneous. One of the intermediates is a carbonium ion. Note that the whole process is reversible.

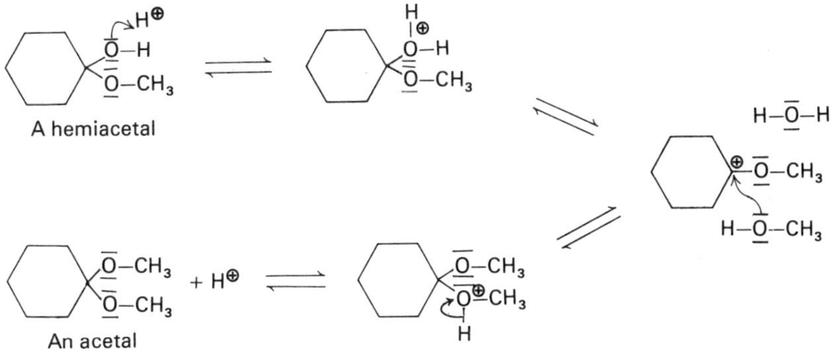

i.e.

$$R_2C{=}O + 2R'OH \rightleftharpoons R_2C(OR')_2 + H_2O$$

The formation of acetals from aldehydes or ketones and alcohols is a general reaction, although the value of the equilibrium constant varies with the nature of the groups R and R'.

Problem 15–4. What would be formed if the cyclic ether shown below were mixed with dilute HCl? How do you think the ether could be made?

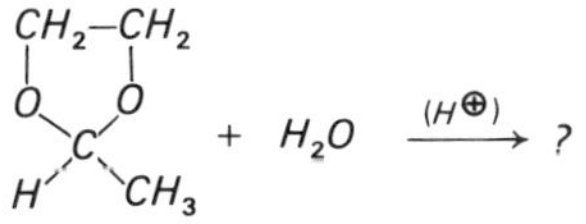

## $H^\oplus$ + amines

Probably the most useful reaction of protonated carbonyl groups is with amines. We shall discuss a general primary amino compound $R{-}NH_2$.

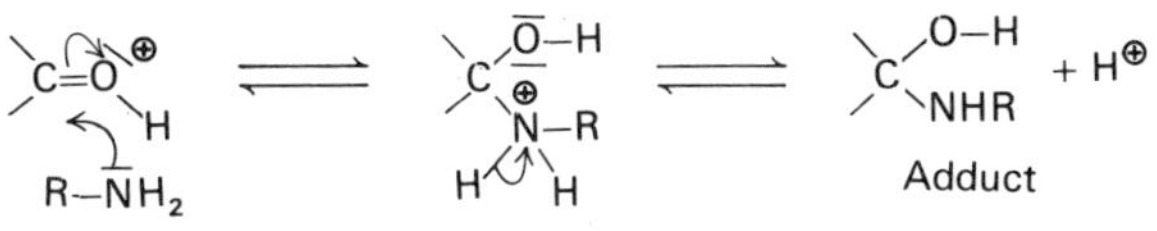

The reaction starts in the same way as the reaction with alcohols. The product at this stage is a 1 : 1 adduct of carbonyl compound plus amine. In many cases this is not isolatable. Further protonation and loss of water can give a delocalized cation, as happens also during acetal formation:

$$\underset{\diagdown}{\overset{\diagup}{C}}\underset{NHR}{\overset{\overset{\oplus}{O}H_2}{}} \rightleftharpoons \overset{\oplus}{C}-\overset{-}{\underset{\underset{H}{|}}{N}}R + H_2O$$

This could react with a second molecule of the amine to give a product, $C(NHR)_2$, analogous to an acetal, or it could lose a proton from the nitrogen to give an *imine*:

$$\underset{R'}{\overset{R}{}}\overset{\oplus}{C}-\overset{-}{\underset{\underset{H}{|}}{N}}-R \rightleftharpoons \underset{R'}{\overset{R}{}}C=\overset{-}{N}-R + H^{\oplus}$$
An imine

This is possible only if a primary amine is used initially. If a secondary amine is used initially there will be no proton on the nitrogen at this stage, but there might be one on the adjacent carbon atom:

$$\underset{H}{\overset{H}{}}\overset{H}{\underset{\underset{H}{|}}{C}}-\overset{\oplus}{\underset{\underset{H}{|}}{C}}-\overset{R}{\underset{}{N}}-R \longrightarrow \underset{H}{\overset{H}{}}C=\overset{}{\underset{\underset{H}{|}}{C}}-\overset{-}{N}R_2 + H^{\oplus}$$
An enamine

Loss of a proton from carbon would give an aminoalkene or *enamine*.

The three reactions can be summarized as follows:

$$R_2C-\underset{\underset{H}{|}}{\overset{\overset{O}{\diagup}}{C}}\diagdown R
\begin{cases}
+\ 2NHR_2 \xrightleftharpoons{(H^{\oplus})} R_2CH-\underset{\underset{R}{|}}{C}\underset{NR_2}{\overset{NR_2}{<}} + H_2O \\
\qquad\qquad\qquad\qquad\text{Diamine} \\
+\ NH_2R \xrightleftharpoons{(H^{\oplus})} R_2CH-\underset{\underset{R}{|}}{C}=NR + H_2O \\
\qquad\qquad\qquad\qquad\text{Imine} \\
+\ NHR_2 \xrightleftharpoons{(H^{\oplus})} R_2C=\underset{\underset{R}{|}}{C}-NR_2 + H_2O \\
\qquad\qquad\qquad\qquad\text{Enamine}
\end{cases}$$

They are all reversible.

The course the reaction takes depends on the structures of the amine and of the carbonyl compound. Imine formation is usual with primary amines and enamine formation with secondary amines.

*Problem 15–5. Why do secondary amines not condense with ketones to give imines?*

*Problem 15–6. Predict whether the equilibrium constant for the following reaction is greater or less than 1; remember that primary amines do not give enamines and that the reactions above are all under equilibrium control:*

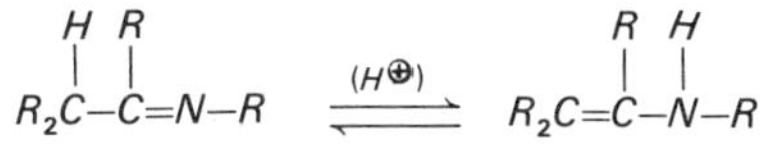

$$R_2\overset{\overset{\displaystyle H}{|}}{C}-\overset{\overset{\displaystyle R}{|}}{C}=N-R \quad \underset{\displaystyle \rightleftharpoons}{\overset{(H^{\oplus})}{\longrightarrow}} \quad R_2C=\overset{\overset{\displaystyle R}{|}}{C}-\overset{\overset{\displaystyle H}{|}}{N}-R$$

Aldehydes and ketones both react with primary amines such as hydroxylamine, $NH_2OH$, to give imines. The hydroxyimines formed in this particular case are called *oximes*:

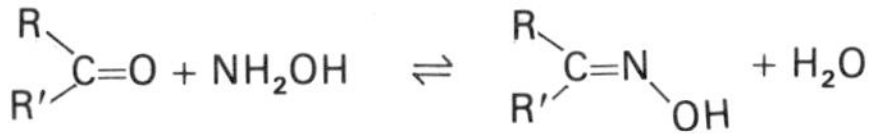

$$\underset{R'}{\overset{R}{>}}C=O + NH_2OH \quad \rightleftharpoons \quad \underset{R'}{\overset{R}{>}}C=N\underset{OH}{} + H_2O$$

An oxime

Oximes are crystalline solids. The melting points of the oximes can be used to help to identify the (usually liquid) carbonyl compounds.

*Problem 15–7. Draw out the structures of the products to be expected from analogous reactions between the ketone RCOR' and 2,4-dinitrophenylhydrazine (see p. 129), and between the aldehyde RCHO and aniline.*

*Problem 15–8. When you drew the reaction with the dinitrophenylhydrazine did you stop to think why the primary rather than the secondary amino group reacts as the nucleophile? What effect will the dinitrophenyl group have on the ability of the secondary amino group to donate electrons? Is this in accord with the discussion of the basicity of aniline on p. 131? Will 2,4-dinitroaniline be more or less basic than aniline?*

*Problem 15–9. Go back to the beginning of the discussion of the reaction of amines with protonated ketones. Have we forgotten something? What happens when amines are added to protonated alcohols. See p. 98. Which is more acidic, a protonated alcohol or a protonated ketone? Why?*

## SUMMARY

Carbonyl groups readily undergo addition reactions. Nucleophiles add to the carbon atom and electrophiles to the oxygen atom. The initial adducts are not always the final products. A wide variety of groups can be added across the C=O double bond so that ketones and aldehydes are a useful starting point for making other types of molecule. Only a few of the substances which add to C=C double bonds can be added to C=O double bonds and only a few of the substances which add to C=O double bonds can be added to C=C double bonds. The synthetic potential of carbonyl groups is illustrated in the following problems.

*Problem 15–10. What products would you expect from the following reactions?*

(a)  [cyclopentanone structure] + *NaBH₄ in H₂O;*

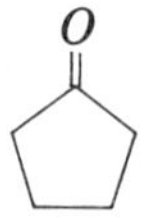

(b)  [cyclopentanone structure] + *CH₃CH₂MgBr then dilute HCl;*

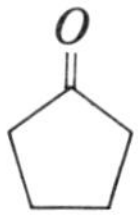

(c)  [cyclopentanone structure] + *excess of methanol containing a small amount of H₂SO₄;*

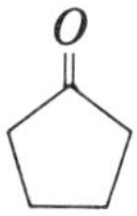

(d)  [cyclopentanone structure] + *NH₂OH + a trace of acid;*

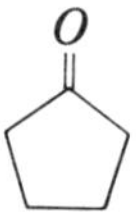

(e)  [cyclopentanone structure] + [pyrrolidine structure, N–H] + *a trace of acid;*

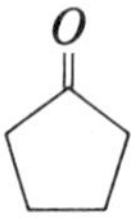

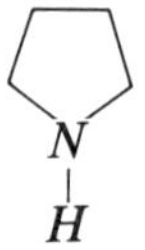

(f)  [cyclopentane with CH₃O OCH₃ substituent] + *dinitrophenylhydrazine + H⊕.*

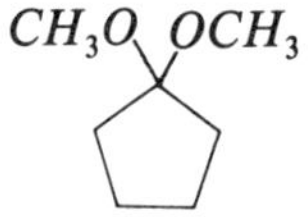

*Problem 15–11. How could you convert* [CH₃CH(CH₃)CHO structure] *into* [isopropyl phenyl ketone structure] *?*

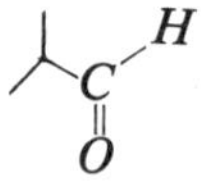

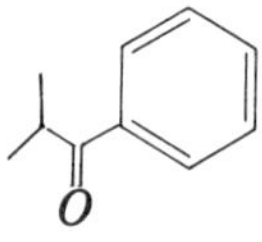

## ANSWERS TO PROBLEMS IN CHAPTER 15

*Problem 15–1.*

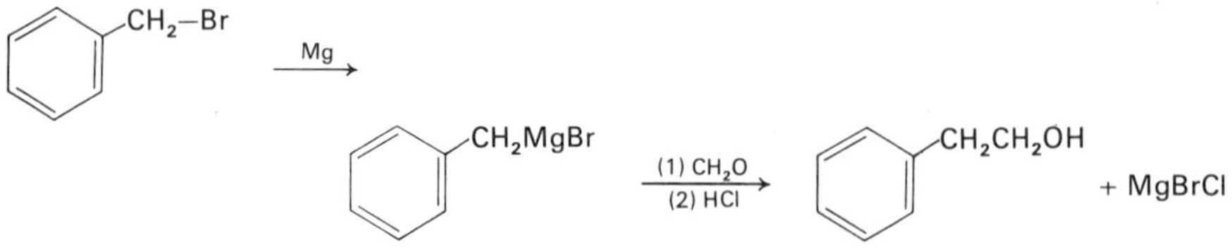

[benzyl bromide C₆H₅CH₂–Br] $\xrightarrow{Mg}$ [C₆H₅CH₂MgBr] $\xrightarrow[(2) HCl]{(1) CH₂O}$ [C₆H₅CH₂CH₂OH] + MgBrCl

*Problem 15–2.* Dissolve the impure cyclohexanol in ether. Shake it with aqueous sodium hydrogen sulphite. The ketone will combine with this reagent and form the bisulphite adduct shown, which will dissolve in the aqueous layer. Separate and dry the ether layer. Distil off the ether to leave pure cyclohexanol.

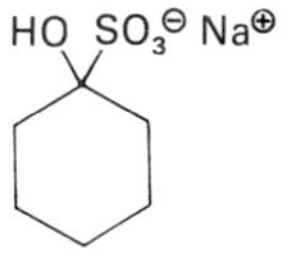

*Problem 15–3.*

$$CH_2=CHCH_2CH_2CH=O \xrightarrow{H_2(Pt)} CH_3CH_2CH_2CH_2CHO$$

Catalytic hydrogenation of the C=C double bond is faster than that of the C=O double bond.

$$CH_2=CHCH_2CH_2CH=O \xrightarrow[\text{in water}]{NaBH_4} CH_2=CHCH_2CH_2CH_2OH$$

$NaBH_4$ and $LiAlH_4$ do not reduce isolated C=C double bonds.

*Problem 15–4.* The cyclic ether is an acetal. On hydrolysis under acid catalysis it gives a carbonyl compound and an alcohol:

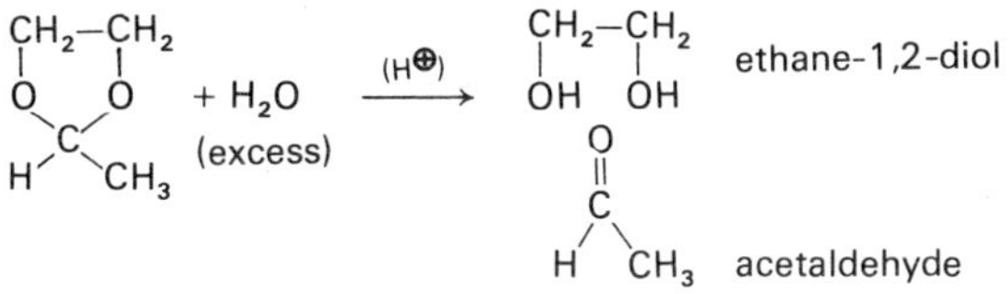

The reaction is reversible. The cyclic acetal could be made from acetaldehyde and ethane-1,2-diol in the presence of an acid if the water formed were removed from the reaction mixture.

*Problem 15–5.* Secondary amines have only one hydrogen attached to nitrogen, so cannot form imines by reaction with ketones.

*Problem 15–6.* The imine and the isomeric enamine are interconvertible in the presence of acid:

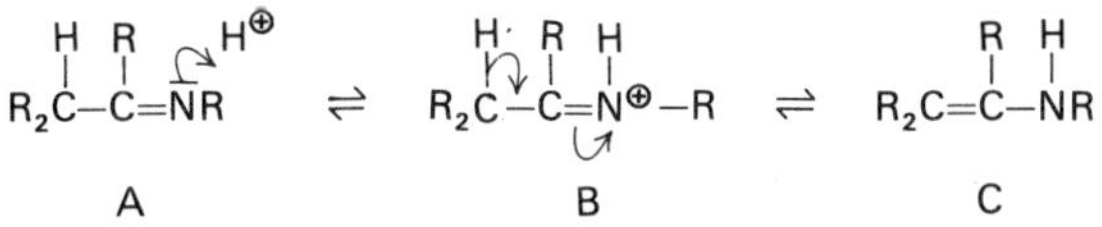

We know that when a delocalized cation such as B is formed from a ketone and a primary amine, it decomposes to a proton and an imine. We must conclude that the formation of A or C from B is under thermodynamic control and that A is more stable than C. In other words, the equilibrium constant, $K = \dfrac{[C]}{[A]}$, for the interconversion A ⇌ C is less than 1.

*Problem 15–7.*

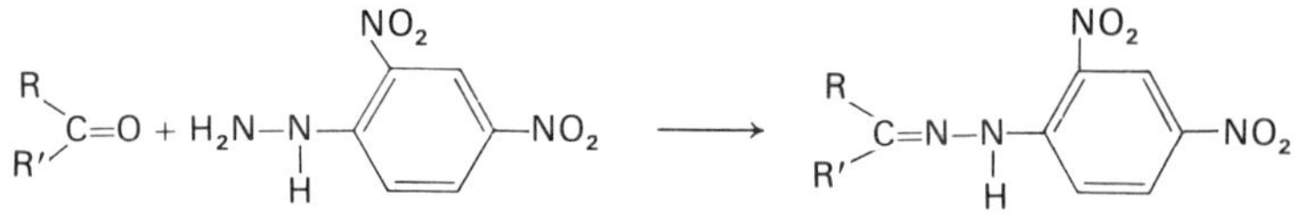

$$\begin{matrix} R \\ \diagdown \\ \diagup \\ R' \end{matrix} C{=}O + H_2N{-}N{-}\!\!\left\langle \text{(ArNO}_2)_2 \right\rangle \longrightarrow \begin{matrix} R \\ \diagdown \\ \diagup \\ R' \end{matrix} C{=}N{-}N{-}\!\!\left\langle \text{(ArNO}_2)_2 \right\rangle$$

A dinitrophenylhydrazone

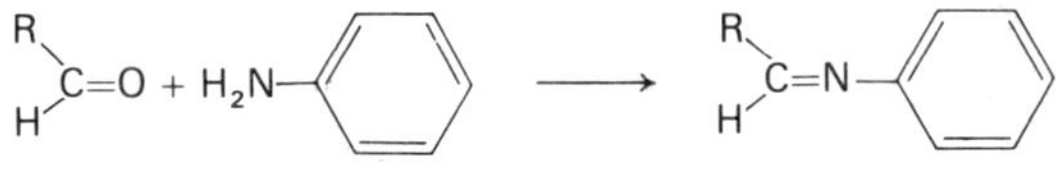

$$\begin{matrix} R \\ \diagdown \\ \diagup \\ H \end{matrix} C{=}O + H_2N{-}\!\!\left\langle \text{Ph} \right\rangle \longrightarrow \begin{matrix} R \\ \diagdown \\ \diagup \\ H \end{matrix} C{=}N{-}\!\!\left\langle \text{Ph} \right\rangle$$

*Problem 15–8.* As we saw in Chapter 13, the lone pair of electrons on the nitrogen atom of aniline is involved in the delocalization and is not so readily available for bond formation as the lone pair of electrons in methylamine. Since the secondary amino group in dinitrophenylhydrazine is like the amino group in aniline whereas the primary one is not, the primary amino group is the stronger base and the better nucleophile. The delocalization in dinitroaniline and dinitrophenylhydrazine is even more extensive than in aniline and dinitroaniline is a *very* weak base. Both nitro groups withdraw electrons from the amino group.

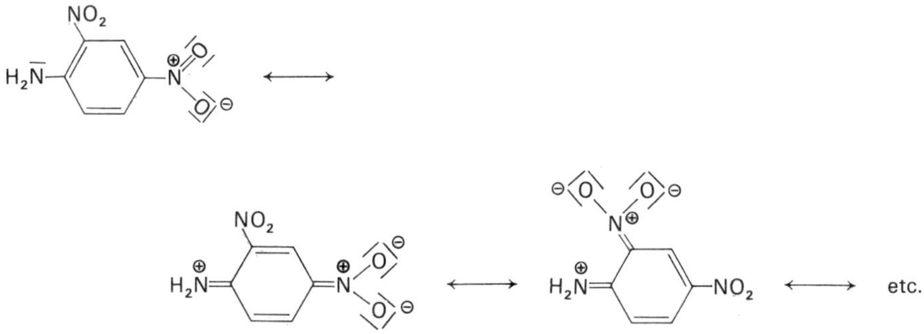

*Problem 15–9.* We must ask why the amine does not simply remove the proton from the protonated ketone and so stop the reaction. The answer is that an equilibrium:

$$R_2C{=}\overset{\oplus}{O}{-}H + NH_2R \rightleftharpoons R_2C{=}O + {}^{\oplus}NH_3R$$

is set up and is much less one-sided, because of delocalization of the charge on the oxonium ion, than the analogous equilibrium:

$$R{-}\overset{\oplus}{O}H_2 + NH_2R \rightleftharpoons ROH + {}^{\oplus}NH_3R$$

When a *small amount* of acid is added to a mixture of an amine and a ketone, some of the amine is converted into an ammonium ion but some amine and some protonated ketone are always present and they can react rapidly to give the observed products.

*Problem 15–10.*

(a) cyclopentanone + NaBH$_4$ in water ⟶ Cyclopentanol

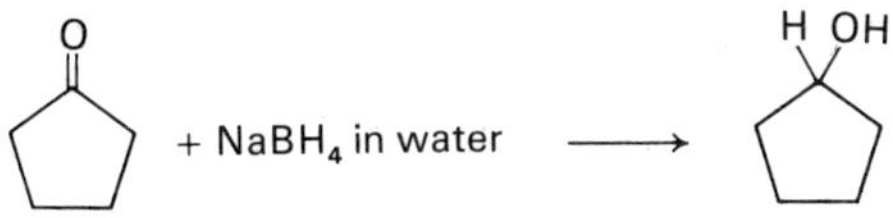

(b) cyclopentanone + CH$_3$CH$_2$MgBr then H$^\oplus$ ⟶ 1-Ethylcyclopentanol

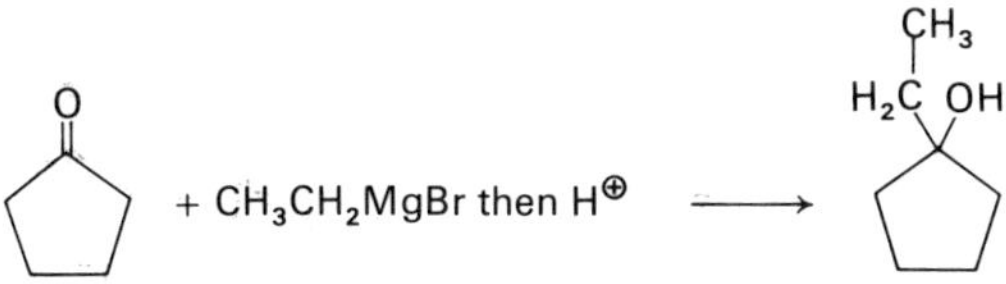

(c) cyclopentanone + CH$_3$OH $\underset{(H^\oplus)}{\rightleftharpoons}$ (hemiketal CH$_3$O OH) $\rightleftharpoons$ (CH$_3$O OCH$_3$) + H$_2$O

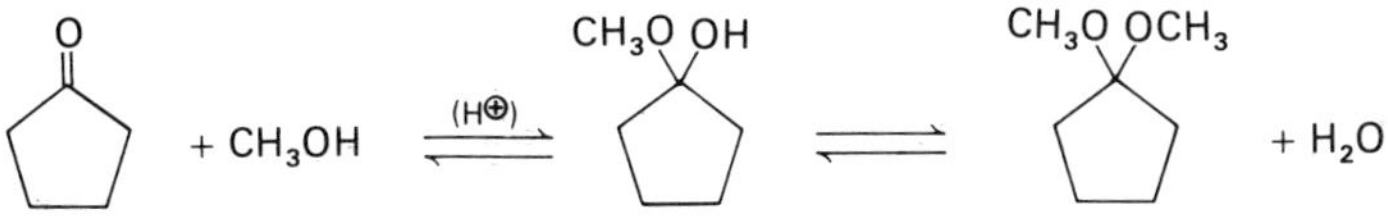

1,1-Dimethoxycyclopentane
(cyclopentanone dimethyl acetal)

If a large excess of methanol were used or if the water were removed, the acetal would be the major product.

(d) cyclopentanone + H$_2$NOH $\xrightarrow{(H^\oplus)}$ Cyclopentanone oxime (N–OH)

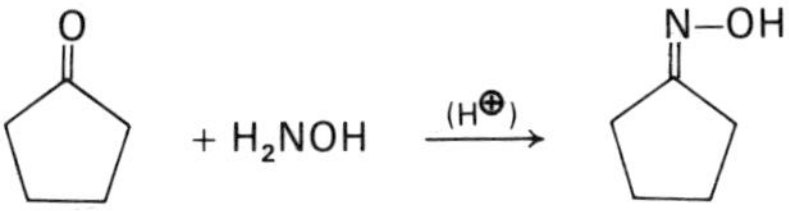

(e) cyclopentanone + pyrrolidine $\xrightarrow{(H^\oplus)}$ An enamine

(f)

(CH$_3$O OCH$_3$) + NH$_2$NH–(2,4-dinitrophenyl) $\xrightarrow{(H^\oplus)}$ 2CH$_3$OH + N=N–NH–(2,4-dinitrophenyl)

↕ H$^\oplus$

CH$_3$–O$^\oplus$O–CH$_3$ (with H) ⇌ CH$_3$–O$^\oplus$ $\xrightarrow{H_2N–R}$ CH$_3$O N$^\oplus$HR $\xrightarrow[\text{(2) loss of CH}_3\text{OH}]{\text{(1) proton transfer}}$ NHR$^\oplus$

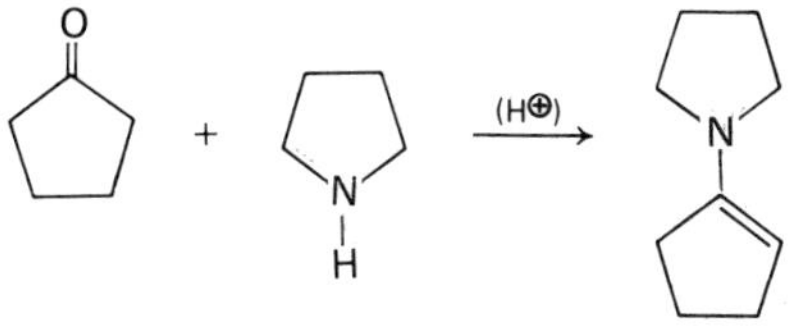

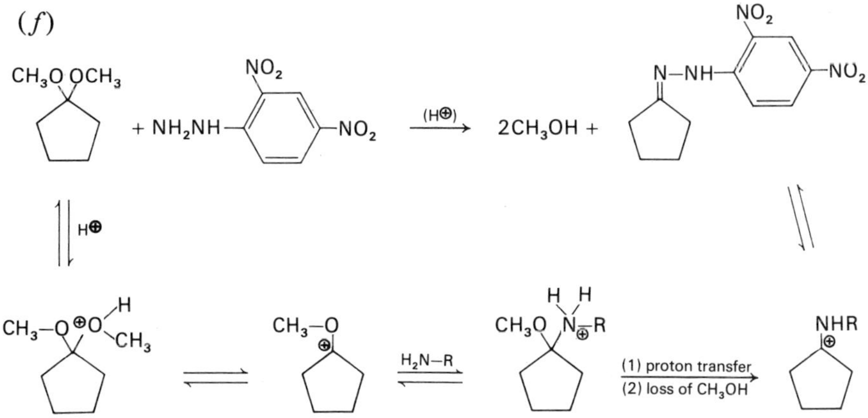

160

The sequence is similar to those involved in the formation of acetals and of imines from ketones. The acetal would be converted into the dinitrophenylhydrazone, which would probably precipitate and so be removed from the equilibrium.

*Problem 15–11.* Direct substitution of H for phenyl is not possible. Indirect routes include

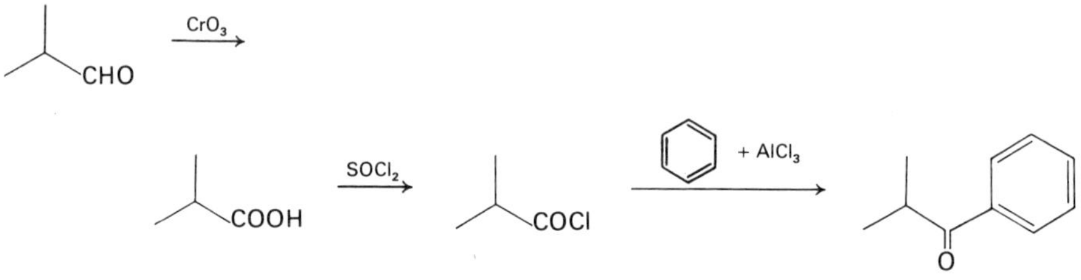

or better

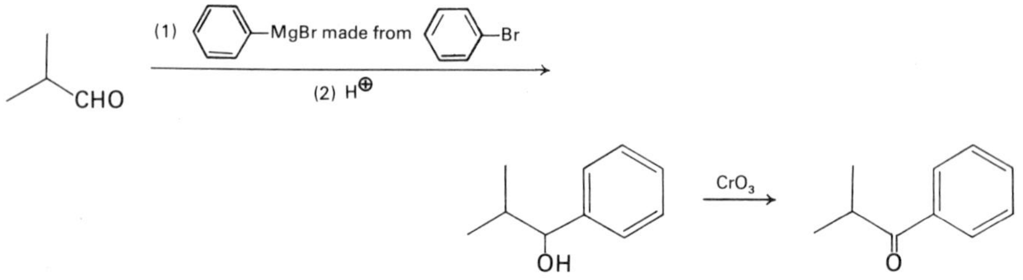

# Chapter 16

## Carboxylic Acids and their Derivatives

The chemistry of alcohols, alkoxide ions, amines, ethers, and alkyl halides has been discussed in previous chapters, and so has the chemistry of carbonyl groups. In the series of compound types below, these functional groups are immediately adjacent and affect one another's behaviour. These types of compound are important and have names of their own.

$$\begin{array}{ccc} \overset{\displaystyle O}{\underset{\displaystyle \|}{R-C-OH}} & \overset{\displaystyle O}{\underset{\displaystyle \|}{R-C-O^{\ominus}}} & \overset{\displaystyle O}{\underset{\displaystyle \|}{R-C-NH_2}} \\ \text{A carboxylic} & \text{A carboxylate} & \text{An amide} \\ \text{acid} & \text{ion} & \end{array}$$

$$\begin{array}{ccc} \overset{\displaystyle O}{\underset{\displaystyle \|}{R-C-OR'}} & \overset{\displaystyle O}{\underset{\displaystyle \|}{R-C-Cl}} & \overset{\displaystyle O\quad O}{\underset{\displaystyle \|\quad\|}{R-C-O-C-R}} \\ \text{An ester} & \text{An acid chloride} & \text{An acid anhydride} \end{array}$$

The members of this series are structurally related. They can all be made from the first one, the carboxylic acid, and can all be re-converted to the acid fairly easily.

The names of individual members of each of these types of compound are derived from the names of the acids. Many acids that have been known for a long time have trivial names which are still in use. The systematic names end in -oic acid. The names of some acids are listed in the table overleaf.

The carbon chain of acids is always numbered starting with the carbon of the carboxyl group as C-1, as in the following examples:

$$\begin{array}{l} \overset{\displaystyle CH_3}{\underset{\displaystyle CH_3}{\diagdown}}C=CHCOOH \qquad \text{3-methylbut-2-enoic acid} \\ \nearrow \\ \end{array}$$

$$\underset{\displaystyle OH}{\overset{\displaystyle}{CH_3-CH-COOH}} \qquad \text{2-hydroxypropanoic acid (lactic acid)}$$

161

| Formula | Systematic name | Trivial name | B.p. (°C) | Solubility in water |
|---|---|---|---|---|
| $H-\overset{\displaystyle O}{\overset{\displaystyle \|}{C}}-O-H$ or $HCO_2H$ | Methanoic acid | Formic acid | 101 | Soluble |
| $CH_3COOH$ or $HOCOCH_3$ or $CH_3CO_2H$ | Ethanoic acid | Acetic acid | 118 | Soluble |
| $CH_3CH_2COOH$ | Propanoic acid | Propionic acid | 141 | Soluble |
| $CH_3CH_2CH_2COOH$ | Butanoic acid | Butyric acid | 163 | Soluble |

In the case of aromatic carboxylic acids related to benzoic acid, the carbon atoms of the ring are numbered starting with that carrying the carboxyl group.

| Formula | Systematic name | Trivial name | M.p. (°C) | Solubility in water |
|---|---|---|---|---|
| 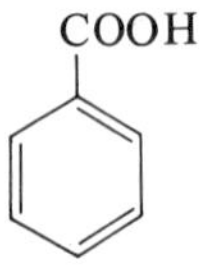 COOH (benzene ring) | Benzoic acid | — | 122 | Insoluble |
| 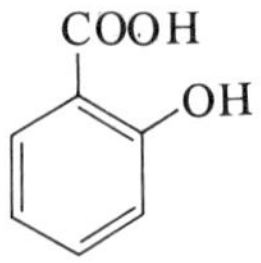 COOH, OH (benzene ring) | 2-Hydroxybenzoic acid (o-hydroxybenzoic acid) | Salicylic acid | 159 | Insoluble |
| 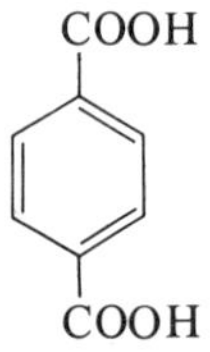 COOH, COOH (benzene ring) | Benzene-1,4-dicarboxylic acid | Terephthalic acid | — | Insoluble |

Carboxylic acids are so called because they are acidic, that is they donate protons to water and they are deprotonated totally by bases (e.g. NaOH) giving salts (sodium carboxylates). If the base used is sodium carbonate, then carbonic acid, $H_2CO_3$ is formed. This is structurally related to carboxylic acids, $RCO_2H$, but it decomposes, giving bubbles of carbon dioxide, and this is a good test for the presence of a $-COOH$ group.

$$2RCOOH + Na_2CO_3 \rightarrow 2RCOO^{\ominus}\,Na^{\oplus} + H-O-\overset{\displaystyle O}{\overset{\displaystyle \|}{C}}-O-H \rightarrow H_2O + CO_2$$
$$\text{Carbonic acid}$$

They also donate protons to indicators such as litmus, which are weak bases, and thereby change the colour of the indicator (in the case of litmus from blue to red). A more quantitative discussion of their acidity and an interpretation of the cause of it are given below. Carboxylic acids are rather polar and form hydrogen bonds via the —O—H and the =O groups. They are fairly high boiling liquids or crystalline solids. The smaller, more volatile acids have piquant or nauseating smells. Those with small hydrocarbon groups are soluble in water. Most are soluble in organic solvents, including non-polar solvents such as benzene in which the acids form hydrogen bonds to other molecules of themselves in pairs.

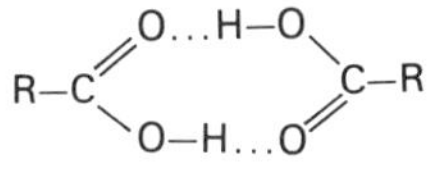

$$R-C \overset{O \cdots H-O}{\underset{O-H \cdots O}{\diagup}} C-R$$

As already mentioned, the naming of derivatives is based on the name of the acid. For example, acid chlorides are named as follows:

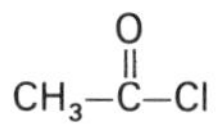

$$CH_3-\overset{\overset{O}{\|}}{C}-Cl \qquad CH_3CH_2CH_2\overset{\overset{O}{\|}}{C}-Cl$$

Acetyl chloride          Butanoyl chloride 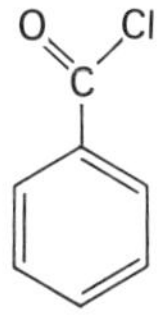          Benzoyl chloride
(ethanoyl chloride)

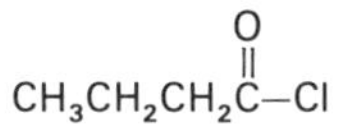

Any group of the type $R-C\overset{O}{\diagdown}$ is called an acyl group. The group $CH_3-C\overset{O}{\diagdown}$ is called the acetyl (ethanoyl) group since it is present in acetic acid (ethanoic acid) and its derivatives. Anhydrides contain two acyl groups. They may be symmetrical as in acetic (ethanoic) anhydride:

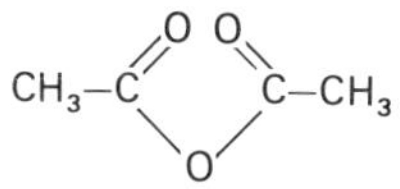

$$CH_3-C\overset{\diagup O}{\underset{\diagdown O \diagup}{}} \overset{O\diagdown}{} C-CH_3$$

Acetic anhydride

or unsymmetrical, RCOOCOR′.
Salts are named as in

$$CH_3CH_2CO_2^{\ominus} Na^{\oplus}$$

Sodium propanoate
(sodium propionate)

The esters are named as in

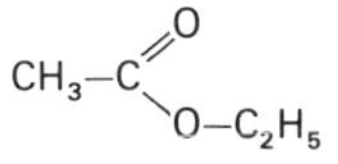

$$CH_3-C\overset{\diagup O}{\underset{\diagdown O-C_2H_5}{}} \qquad CH_3-CH_2-C\overset{\diagup O}{\underset{\diagdown O-CH_3}{}}$$

Ethyl acetate          Methyl propionate
or ethyl ethanoate     or methyl propanoate

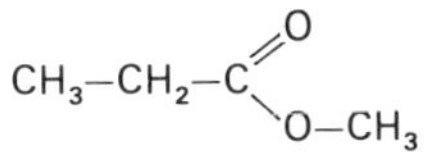

The amides are named in various ways:

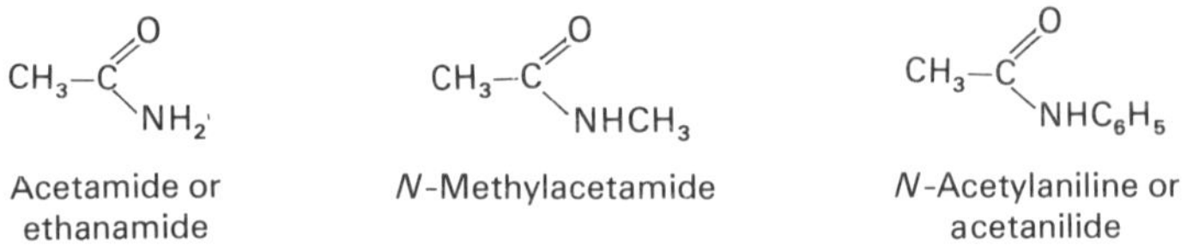

| Acetamide or ethanamide | N-Methylacetamide | N-Acetylaniline or acetanilide |

As before (Chapter 12), the prefix N- indicates that the substituent is attached to nitrogen.

*Problem 16–1. Write the structural formulae for ethyl propanoate, benzamide (the amide of benzoic acid), propanoyl chloride, dimethyl carbonate, sodium hydrogen carbonate, acetic anhydride, and 4-acetylbenzoic acid.*

The very fact that each of these compound *types* has a name of its own suggests that the carboxyl group or the amide group shows a chemistry distinct from that of compounds with an isolated carbonyl, hydroxyl, or amino group. The reactions of the amino group and of the carbonyl group are in fact much modified when they are adjacent. For example, the amino group in amides is non-basic and poorly nucleophilic and the carbonyl group of amides is a poorer electrophile than the carbonyl group of ketones.

The discussion of substituted alkenes and substituted benzenes above showed that reactions of a functional group can be affected by the electronegativity of atoms attached to it. A much more important interaction occurs if the electrons in the adjacent groups are delocalized. This is the effect which dominates the chemistry of all the derivatives of carboxylic acids.

## CARBOXYLATE IONS

Carboxylate ions are present in salts such $RCO_2^{\ominus} Na^{\oplus}$ or $(RCO_2^{\ominus})_2 Mg^{2+}$, which are crystalline solids soluble in water. The carboxylate ion has been represented as

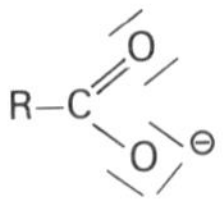

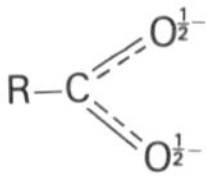

The true electron distribution, as deduced from spectral data, is such that the two C–O bonds are identical and the oxygen atoms each carry half of the negative charge. We can draw this symmetrical distribution as

or, by using two pictures, as

Because of this delocalization of the charge over two oxygen atoms, the ion is very stable. It is a much poorer nucleophile and a much weaker base than an alkoxide ion. Also because of the delocalization, the carbon atom of the carbonyl group does not accept electrons from nucleophiles nearly so readily as does the carbonyl carbon of ketones. Ketones react with cyanide ion and with tetrahydroborate ion. Carboxylate ions do not:

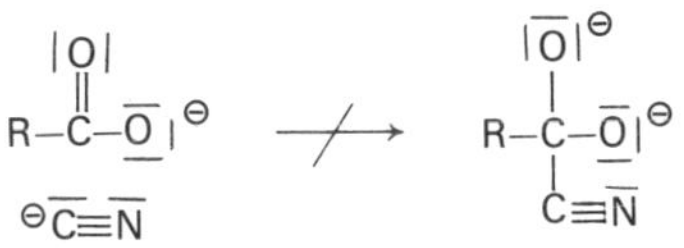

Thus, if delocalization becomes possible or more extensive when two functional groups are adjacent, the chemistry of each will be affected, the new molecule will be particularly stable, and its shape may be more rigidly controlled (see p. 58).

*Problem 16–2. What product would you expect from the reaction of the salt of the keto acid shown below with sodium borohydride in water?*

$$CH_3-\overset{O}{\underset{\|}{C}}-CH_2-CH_2-\overset{O}{\underset{\|}{C}}-O^{\ominus} Na^{\oplus} + NaBH_4 \rightarrow ?$$

Delocalization is complete in the carboxylate ion in the sense that the true electron distribution is half way between the two pictures which can be drawn. The two oxygen atoms are equivalent.

In the other derivatives of acids listed on p. 161, the carbonyl group is less able to withdraw electrons from the adjacent functional group, whose lone pair(s) is thus more localized than in carboxylate ions.

## AMIDES

The electrons in amides are delocalized. The true situation is

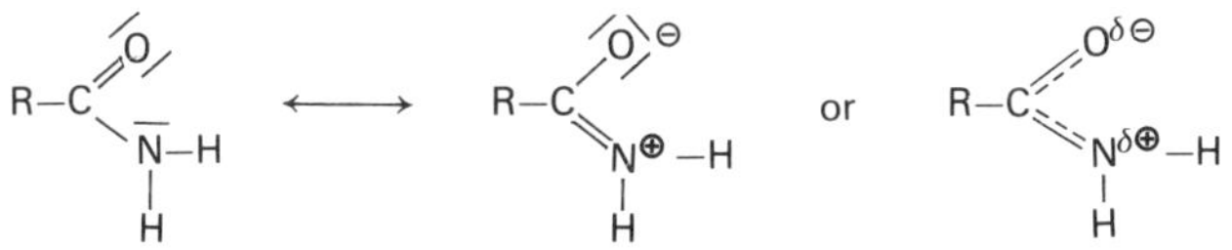

Amides are therefore rather polar molecules with a positive end and a negative end. The polar nature of the molecules and the possibility of hydrogen bonding

mean that there are strong attractive forces between amide molecules and most amides are solids.

The carbonyl group is not readily attacked by nucleophiles and the nitrogen atom does not readily donate electrons to electrophilic reagents. Amides, unlike amines, are neutral. Because of the partial double bond character of the C$-$N bond, the $\overset{\displaystyle O}{\diagdown}$ $C-N\diagup$ part of the molecule is planar with 120° bond angles and rotation about the C$-$N bond is difficult.

## ESTERS

In carboxylic acids and esters the oxygen atom of the OH or OR group is not so willing to lose electrons to the carbonyl group as was the nitrogen of amides or the oxygen of the carboxylate ion. Delocalization in these molecules is slight; that is, the first drawing for the electron distribution is not very far from the true situation:

$$
R-C\underset{O-H}{\overset{O}{\diagup}} \quad \longleftrightarrow \quad R-C\underset{\overset{\oplus}{O}-H}{\overset{\overset{\ominus}{O}}{\diagup}}
$$

$$
R-C\underset{O-R'}{\overset{O}{\diagup}} \quad \longleftrightarrow \quad R-C\underset{\overset{\oplus}{O}-R'}{\overset{\overset{\ominus}{O}}{\diagup}}
$$

Esters are not very polar and are usually liquids, although of course the nature of the groups R and R$'$ will affect the physical properties. Many esters have pleasant smells and many flowers and fruits owe their characteristic aromas to the presence of esters. Esters are slightly less easily attacked at carbon by nucleophiles than are ketones. For instance, ketones are reduced by sodium tetrahydroborate but most esters are not.

## ACID CHLORIDES

There is no delocalization in acid chlorides. Part of the reason is the electronegativity of chlorine (approximately the same as that of oxygen) and part derives from the fact that the valency electrons of the chlorine, which are in the third valency shell, do not interact very effectively with those of the carbon, which are in the second valency shell. The carbonyl group of acid chlorides is therefore more electrophilic than that of ketones and reactions with nucleophiles, which result in displacement of chloride ion, are very rapid.

*Problem 16–3. Use arguments such as those above to predict whether thiol esters (see p. 84) or phenyl esters or anhydrides will be more reactive towards nucleophiles than methyl esters are.*

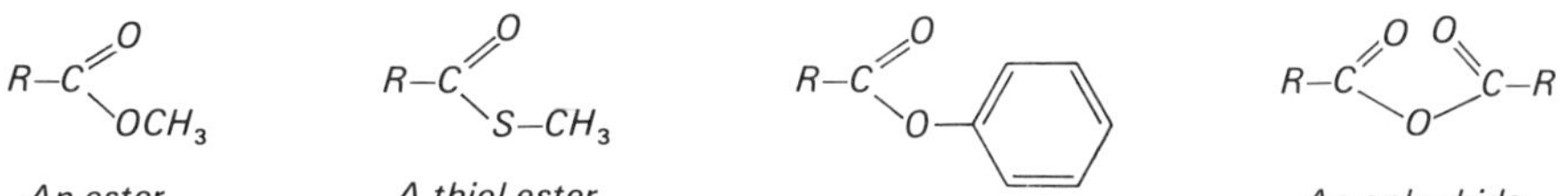

An ester      A thiol ester      A phenyl ester      An anhydride

## REACTIONS OF ACID CHLORIDES AND ANHYDRIDES

Acid chlorides are highly reactive, even towards such weak nucleophiles as water. The reaction of acetyl chloride with water is violently exothermic. The water first adds in a manner typical of carbonyl compounds. The alkoxide ion so produced can, however, decompose with loss of a stable leaving group. The net effect is a substitution in which the carbon is attacked by a *nucleophile* but it is not an $S_N$ substitution like the reaction of methyl chloride with hydroxide ion. This substitution is occurring at a trigonal carbon atom and proceeds in several steps—an addition followed by an elimination and finally a proton transfer to the solvent. Substitutions by addition and then elimination, but involving addition of an *electrophile* to the carbon in the first step, were discussed in Chapter 14 on benzene chemistry.

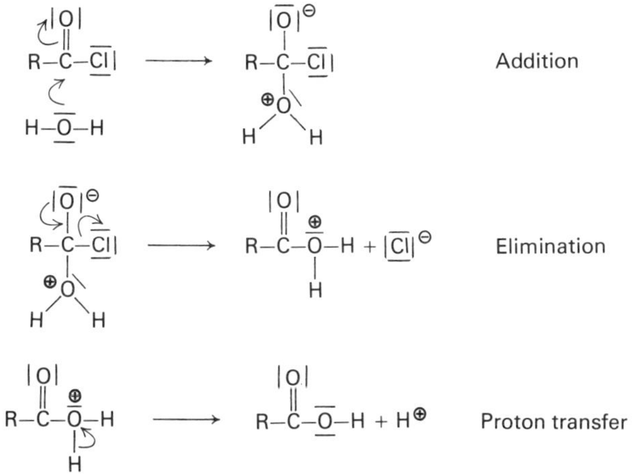

Addition

Elimination

Proton transfer

Overall:

$$RCOCl + H_2O \rightarrow RCOOH + HCl$$

This reaction can be described as the hydrolysis of the acid chloride. Reactions which are of more synthetic value occur between acid chlorides and alcohols or amines or carboxylate ions. The mechanisms of these reactions are analogous to the hydrolysis reaction.

*Problem 16–4. What would be formed in the reaction RCOCl + R'OH?*

*Problem 16–5. How could acetyl chloride be converted into acetanilide (N-phenylethanamide)?*

Anhydrides, like acid chlorides, react vigorously with water:

$$RCO-O-COR + H_2O \rightarrow RCOOH + HOCOR$$

to give two molecules of the acid. This conversion explains why these substances are called the anhydrides of the acids.

*Problem 16–6. What products would you expect from the reaction of acetic anhydride with ethanol?*

As these examples show, the reaction of alcohols or amines with acid chlorides and anhydrides results in replacement of a hydrogen atom by an acyl group. The process is called acylation (or acetylation when the acyl group is $CH_3CO$). Acylation of alcohols and amines is an important route to esters and amides.

## HYDROLYSIS OF ESTERS AND AMIDES

Similar reactions of esters and amides with water would also hydrolyse these to the parent carboxylic acid, plus an alcohol in one case and an amine in the other. However, for the reasons discussed above, these hydrolyses are slower than the hydrolysis of acid chlorides. Esters and amides are unaffected by cold water. There are two ways to make the reaction go faster—to use a better nucleophile such as $^{\ominus}OH$ or to improve the electrophilicity of the carbonyl group by protonating it (see pp. 149 and 152).

In the first case the ester is treated with, say, sodium hydroxide. The products are the salt of the acid plus the alcohol. Hydroxide ions are used up in this process. The base is not a catalyst, but a reagent.

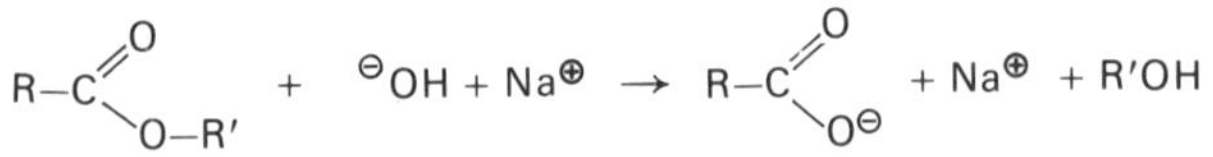

The mechanism is almost the same as that of the hydrolysis of the acid chloride:

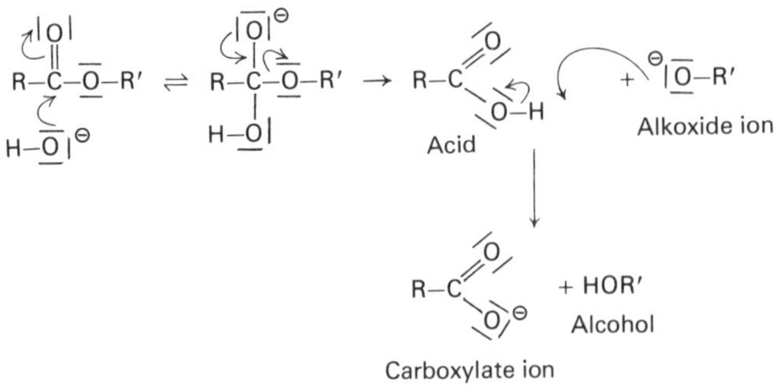

The alkoxide ion is not as good a leaving group as is chloride ion. Once formed, it is strongly basic and captures a proton. The acid, which is formed at the same time, loses a proton in the basic solution. The simplest way to indicate this is to draw the acid losing a proton to the alkoxide ion, although the actual sequence of proton transfers may be more complex and may involve the water.

The product of this alkaline hydrolysis is the salt of the acid. The salt may be converted into the acid itself by adding a strong acid such as HCl:

$$H^{\oplus} + Cl^{\ominus} + RCO_2^{\ominus} + Na^{\oplus} \rightarrow RCO_2H + Na^{\oplus} + Cl^{\ominus}$$

Amides can be hydrolysed by hot aqueous alkali in a process which is essentially the same as the alkaline hydrolysis of esters. A carboxylate ion and ammonia or an amine are formed.

*Problem 16–7. Predict the products from the reaction $CH_3CONHCH_3 + {}^{\ominus}OH$.*

The hydrolysis of esters can also be carried out by using aqueous acid. Some of the protons will become attached to the carbonyl oxygen atom of the ester. This protonated ester is a rather stable carbonium ion but is a better electrophile than the ester itself and will react with water.

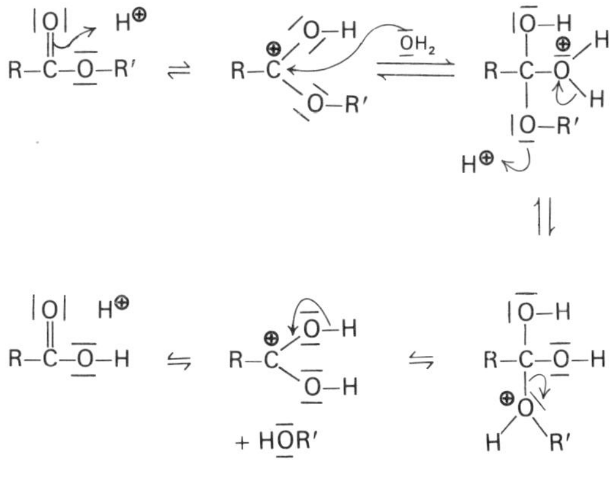

A proton transfer allows the tetrahedral intermediate to break down with loss of alcohol and then loss of a proton to give the carboxylic acid. Overall:

$$RCOOR' + H_2O \xrightarrow{(H^{\oplus})} RCOOH + HOR'$$

The inorganic acid used to bring about this hydrolysis is a true catalyst. It is not consumed in the process.

All of the steps in the above reaction are easily reversible. It is possible, by reversing the whole mechanism, to esterify an acid, i.e. to convert it into an ester, by treating it with the alcohol plus protons as catalyst. An equilibrium will eventually be established whether one starts with acid + alcohol or with ester + water. To convert an acid into an ester using acid catalysis, it is therefore

170

necessary to use a large excess of the alcohol or else to remove the water as it is formed. Likewise, to hydrolyse an ester it is necessary to use an excess of water or (less easily) to remove the alcohol or the acid as it is formed.

*Problem 16–8. In the light of the above discussion, suggest a method of converting an expensive alcohol completely into its acetate.*

*Problem 16–9. Suggest a method of converting the amide below into cyclohexylamine:*

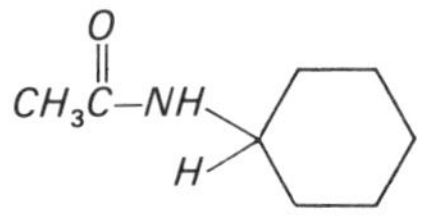

$$CH_3\overset{\overset{\displaystyle O}{\|}}{C}-NH-\hspace{-6pt}\bigcirc$$

*Problem 16–10. What would be the simplest way of converting methyl benzoate into benzamide?*

*Problem 16–11. What would happen if methyl benzoate were heated with a solution of sodium ethoxide in ethanol?*

## MAKING ACID CHLORIDES

So far we have seen that it is possible to convert almost any member of the series on p. 161 into any other one. An important omission is the conversion of acids into acid chlorides. Acid chlorides are very useful reagents because they are so reactive. To make them from the stable acids, the acid is usually treated with a reactive inorganic acid chloride in a non-nucleophilic solvent. One which is convenient is thionyl chloride, $SOCl_2$, which is the double acid chloride of sulphurous acid. It reacts with carboxylic acids to give sulphur dioxide and hydrogen chloride, which escape as gases:

$$RCOOH + SOCl_2 \rightarrow RCOCl + HCl + SO_2$$

The reaction is usually performed in benzene, which can be distilled off together with any unreacted thionyl chloride, leaving only the acid chloride.

## MAKING ESTERS, AMIDES, AND ANHYDRIDES

Since acid chlorides are easily converted into esters, amides, and anhydrides (see p. 168), we now have a good route to these substances from acids:

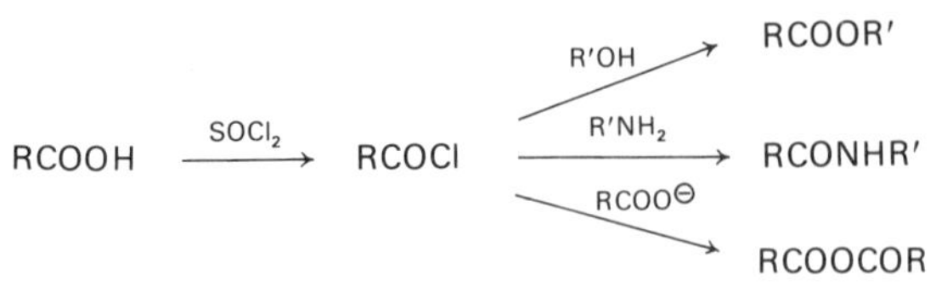

$$RCOOH \xrightarrow{SOCl_2} RCOCl \begin{array}{c} \xrightarrow{R'OH} RCOOR' \\ \xrightarrow{R'NH_2} RCONHR' \\ \xrightarrow{RCOO^\ominus} RCOOCOR \end{array}$$

In the case of esters, this provides an attractive alternative to acid-catalysed esterification of the acid:

$$RCOOH + R'OH \underset{}{\overset{(H^{\oplus})}{\rightleftharpoons}} RCOOR' + H_2O$$

which gives a high yield of ester only if one or other of the reagents is present in excess (see p. 7).

Esters and amides of acetic acid are often made from acetic anhydride, which is commercially available and reacts with alcohols, phenols, and amines rapidly (see p. 168).

## ACIDS AS PROTON SOURCES

Carboxylic acids are acidic, that is, when they are added to water the concentration of hydrogen ions, $[H^{\oplus}]$, increases. So are other compounds such as hydrochloric acid, sulphonic acids ($RSO_3H$), ammonium ions, and phenols. All of these compounds, then, are better proton donors than water. Some other compounds such as alcohols are able to donate protons to bases stronger than water, for example to $HO^{\ominus}$, $H_2N^{\ominus}$, or $H_3C^{\ominus}$. Other compounds such as $CH_4$ will not transfer protons to any base.

We must consider the factors which make some compounds more acidic than others. $H_2S$ is a stronger acid than $H_2O$, and HF is weaker than HCl, so it appears that the size of the atom to which the proton is bonded is important. The larger the atom the more easily the proton is removed.

If we consider atoms of about the same size, as in $\geq$C—H, $>$N—H, —O—H, and F—H, the acidity increases in the same order as the polarity of the bond. In other words, the acidity increases if the electronegativity of the atom which acquires the bonding electrons is increased.

The presence of a charge on the atom carrying the hydrogen has a marked effect. $HO^{\ominus}$ is a much weaker acid than $H_2O$ and $H_3O^{\oplus}$ is a much stronger acid than $H_2O$.

If we now consider the effect of the next-again atom, i.e. the atom X in a series of compounds XOH such as H—OH, $CH_3$—OH, Cl—OH, HO—OH, we find that water and methanol are considerably weaker acids than HOCl or $H_2O_2$, and this can be attributed to the greater electronegativity of the Cl and O atoms compared with H and C.

However, if the group X is of such a kind that delocalization is possible, the acidity increases markedly. Thus, phenol is more acidic than methanol (see p. 131) since extensive delocalization is possible in the phenoxide ion but not in methoxide ion.

Carboxylic acids are even more acidic since the anion is again delocalized but the charge can be carried jointly by two oxygen atoms rather than oxygen and carbon:

The factors which affect acidity therefore include:
the size of the atom to which the hydrogen is attached;
the electronegativity of the atom to which the hydrogen is attached;
the charge on the atom to which the hydrogen is attached;
the electronegativity of substituents on that atom; and
the possibility of delocalization involving these substituents.

*Problem 16–12. Predict which is the stronger acid in each of the following pairs.*

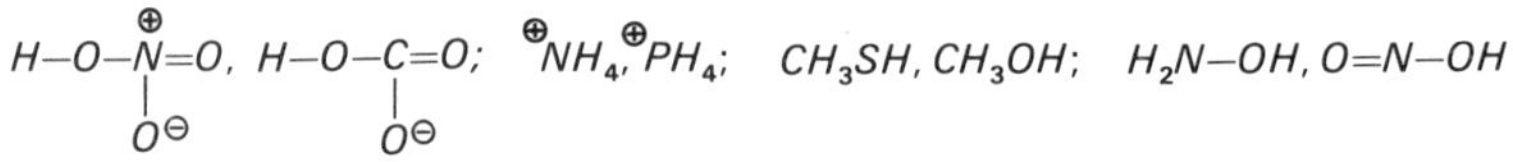

An acid in solution in water is in equilibrium with its anion and a proton:

$$RCOOH \rightleftharpoons RCO_2^{\ominus} + H^{\oplus}$$

Strictly, the proton becomes bonded to a solvent molecule but this does not affect the argument below. As for all chemical equilibria, there is a definite relationship between the concentrations of the products and reagents once equilibrium has been established. In most proton transfers, equilibrium is reached very quickly. The relationship for the acid (at a given temperature for a given acid in a given solvent) can be written as

$$\frac{[RCO_2^{\ominus}][H^{\oplus}]}{[RCOOH]} = constant$$

The concentrations are those of each species actually present in the equilibrium mixture. This relationship holds no matter where the molecules came from. For example, it would hold for the mixture made by adding a small amount of HCl to the salt of the acid. The constant is called the dissociation constant of the acid and given the symbol $K_a$. For acids which dissociate extensively in water, i.e. transfer protons extensively to water, the concentration of anion will be high at equilibrium and the concentration of acid, i.e. covalent RCOOH, will be small. $K_a$ will therefore be large. For weak acids, $K_a$ will be small. A table of $K_a$ values for several acids is given opposite.

The $K_a$ values can be used to predict the equilibrium ratio of two acids and their anions when they are mixed. Thus, phenol is a stronger acid than water. Phenoxide ion is a weaker base than $HO^{\ominus}$. Phenoxide ion does not remove

| Acid | $K_a$ | Degree of dissociation for a 0.1 M solution in water |
|---|---|---|
| HCl | $\sim 10^7$ | $\sim$ complete |
| RSO$_3$H | $\sim 10^3$ | $\sim$ complete |
| RCOOH | $\sim 10^{-5}$ | $\sim$ 1 molecule per 100 |
| H$-$O$-\overset{\displaystyle O}{\overset{\displaystyle \|}{C}}-$O$-$H | $\sim 10^{-7}$ | $\sim$ 1 in 1 000 |
| NH$_4^{\oplus}$ | $\sim 10^{-9}$ | $\sim$ 1 in 10 000 |
| 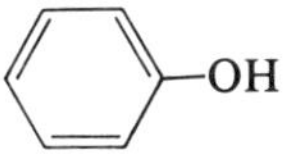 ⬡$-$OH | $\sim 10^{-10}$ | $\sim$ 1 in 30 000 |
| H$-$O$-$H | $\sim 10^{-16}$ | Almost nil |
| C$_2$H$_5-$O$-$H | $\sim 10^{-18}$ | Nil |
| NH$_3$ | Very roughly $10^{-30}$ | Nil |
| CH$_4$ | Very roughly $10^{-40}$ | Nil |

protons from water to any great extent. The equilibrium constant for the reaction below is small:

$$C_6H_5-O^{\ominus} + HOH \; \rightleftharpoons \; C_6H_5OH + HO^{\ominus}$$

Check the cases below by a similar argument.

$$NH_2^{\ominus} + HOH \; \rightleftharpoons \; NH_3 + HO^{\ominus}$$

$$HOCOO^{\ominus} + RCOOH \; \rightleftharpoons \; HOCOOH + RCO_2^{\ominus}$$
$$\downarrow$$
$$H_2O + CO_2$$

$$HOCOO^{\ominus} + C_6H_5OH \; \rightleftharpoons \; HOCOOH + C_6H_5O^{\ominus}$$

Phenoxide ions will accept protons from carbonic acid or stronger acids but not from water, so when sodium phenoxide is dissolved in water it remains largely as the salt, whereas if sodium amide is put into water the anion captures protons to give ammonia and hydroxide ions. Similarly, if the hydrogen carbonate ion is mixed with acetic acid it will capture a proton from this acid to give acetate ion and CO$_2$, since carbonic acid is a weaker acid than acetic acid. However, the hydrogen carbonate ion will not remove protons from phenols, which are weaker acids than carbonic acid.

*Problem 16–13. How would you separate a mixture of phenol and benzoic acid?*

## SUMMARY

The properties of the carboxylic acids and their derivatives can conveniently be summarized in a table, as show below.

| Property | RCOOH acid | RCONH$_2$ amide | RCOOCH$_3$ ester | RCOCl acid chloride | RCO$_2$Na salt |
|---|---|---|---|---|---|
| Polarity | Polar (also hydrogen bonding) | Polar | Moderately polar | Moderately polar | Ionic |
| Physical state: | | | | | |
| when R = CH$_3$ | Liquid | Solid | Liquid | Liquid | Solid |
| when R = C$_6$H$_5$ | Solid | Solid | Liquid | Liquid | Solid |
| Solubility in cold water: | | | | | |
| when R = CH$_3$ | Soluble | Soluble | Insoluble | (Reacts) | Soluble |
| when R = C$_5$H$_5$ | Insoluble | Insoluble | Insoluble | (Reacts) | Soluble |
| Acidity | Acidic | Neutral | Neutral | (Reacts) | Basic |
| Reactions: | | | | | |
| with H$_2$O | None | Very slow | Very slow | Fast | None |
| with HO$^{\ominus}$ | (Proton transfer) | Slow | Moderate | Very fast | None |
| with H$_2$O + H$^{\oplus}$ | None | Slow | Moderate | Very fast | (Gives acid) |

Explanations of these characteristics have been discussed above.

The following problems involve extensions of concepts discussed above.

*Problem 16–14. What ion would be formed in the alkaline hydrolysis of the cyclic ester below?*

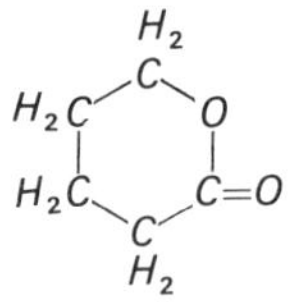

*Problem 16–15. Water-soluble salts of carboxylic acids with long side-chains, e.g. $Na^{\oplus}\ ^{\ominus}OCO(CH_2)_{14}CH_3$, behave as soaps, and esters of such acids occur in vegetable oils, e.g.*

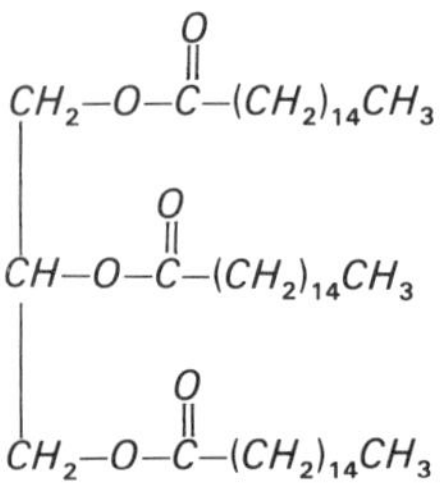

(a component of vegetable oils)

*How could you make some soap if you were wrecked on a desert island?*

*Problem 16–16. Look back at the reaction of LiAlH₄ with ketones (p. 151). LiAlH₄ also reacts with esters. What organic products would you expect from the reaction of methyl butanoate (CH₃CH₂CH₂COOCH₃) with LiAlH₄ followed by water?*

*Problem 16–17. Compounds in which a carbonyl group is attached directly to one other functional group have been discussed above. Extend the arguments presented above to compounds in which two functional groups are attached to a carbonyl group and predict—*

Which of the following is the stronger base: $CH_3-\overset{O}{\underset{\|}{C}}-NH_2$, $NH_2-\overset{O}{\underset{\|}{C}}-NH_2$?

What would be formed on heating $CH_3-O-\overset{O}{\underset{\|}{C}}-O-CH_3$ with aqueous HCl?

How will $Cl-\overset{O}{\underset{\|}{C}}-O-CH_2CH_3$ react with $CH_3NH_2$?
What might happen if you breathed in a small amount of $COCl_2$?

## ANSWERS TO PROBLEMS IN CHAPTER 16

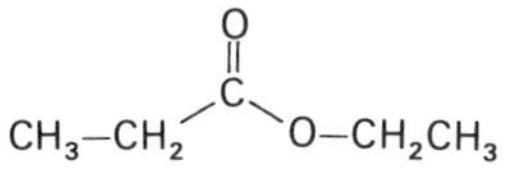

$CH_3-CH_2-\overset{O}{\underset{\|}{C}}-O-CH_2CH_3$

Ethyl propanoate

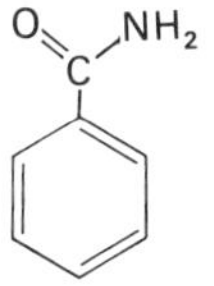

Benzamide

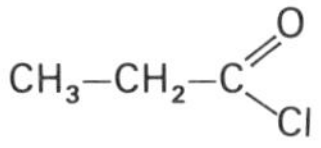

$CH_3-CH_2-C\overset{O}{\underset{Cl}{\diagdown}}$

Propanoyl chloride

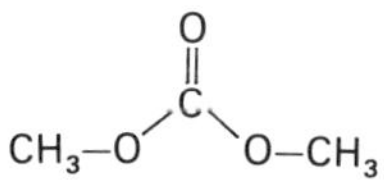

$CH_3-O-\overset{O}{\underset{\|}{C}}-O-CH_3$

Dimethyl carbonate

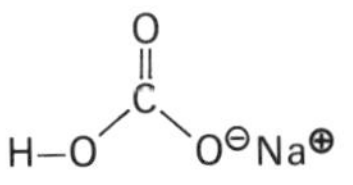

$H-O-\overset{O}{\underset{\|}{C}}-O^{\ominus}Na^{\oplus}$

Sodium hydrogen carbonate
(sodium bicarbonate)

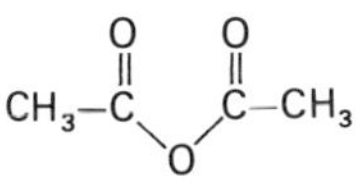

$CH_3-\overset{O}{\underset{\|}{C}}-O-\overset{O}{\underset{\|}{C}}-CH_3$

Acetic anhydride

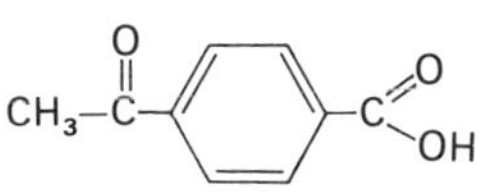

4-Acetylbenzoic acid

*Problem 16–2.* The ketone carbonyl group is sufficiently electrophilic to react with the borohydride ion (p. 151). The carboxylate ion is not, because of the delocalization.

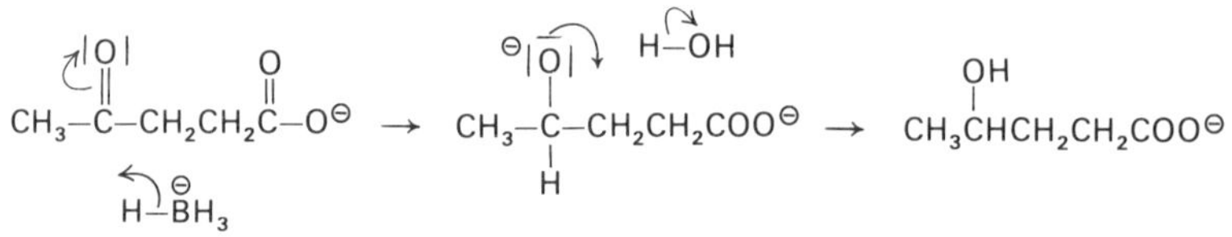

*Problem 16–3.* The reactivity of carbonyl groups towards nucleophiles is reduced if the groups attached to the carbonyl group can release electrons towards it. The most important way in which such release occurs is by delocalization. There will be little delocalization in thiol esters since the lone-pair electrons of sulphur are in the third shell (see p. 166). Thiol esters should therefore be more reactive towards nucleophiles than ordinary esters which are delocalized.

In phenyl esters the lone pair of electrons on oxygen is delocalized over the phenyl group as well as the carbonyl group. Electron release to the carbonyl group is therefore lower than in methyl esters, and phenyl esters should be more reactive towards nucleophiles than methyl esters.

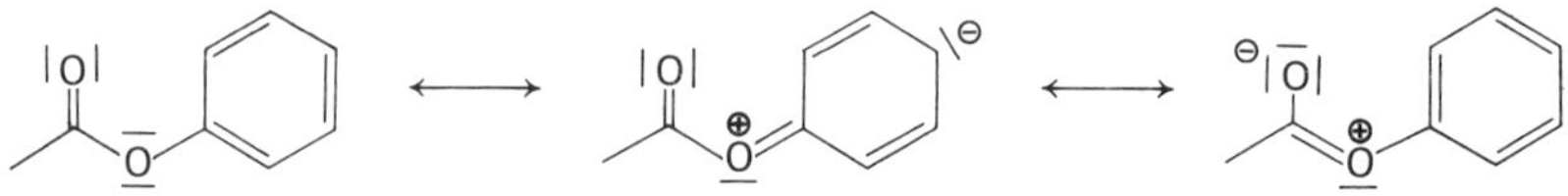

Likewise in anhydrides, delocalization involves the oxygen and *two* carbonyl groups. Both carbonyl groups are therefore more electrophilic than the carbonyl groups of methyl esters.

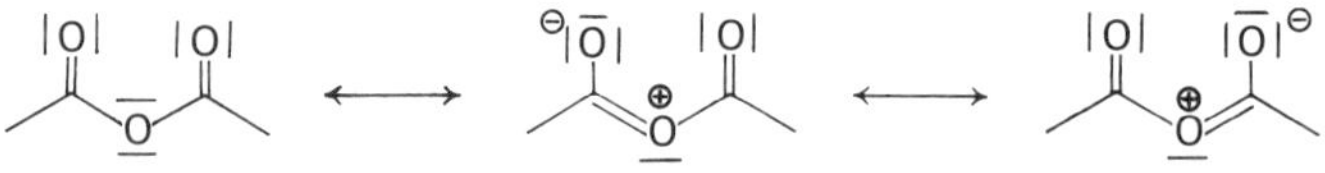

All three of these derivatives are in fact more reactive than methyl esters. Anhydrides are almost as reactive as acid chlorides.

*Problem 16–4.*

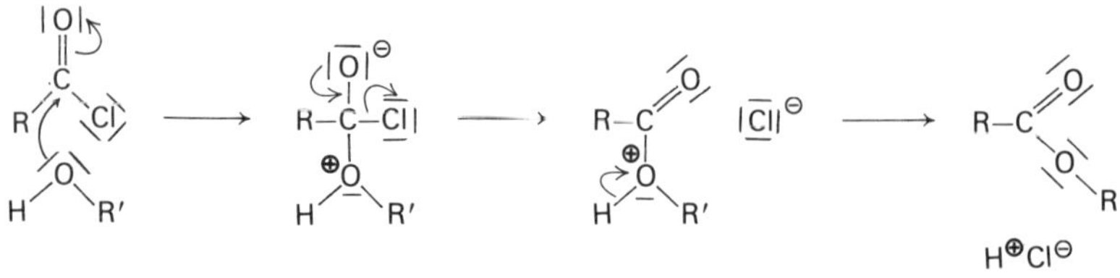

i.e.

$$RCOCl + HOR' \rightarrow RCOOR' + HCl$$

*Problem 16–5.*

$$RCOCl + H_2N-C_6H_5 \rightarrow RCONHC_6H_5 + HCl$$

$$\downarrow H_2NC_6H_5$$

$$\overset{\oplus}{H_3N}-C_6H_5 \quad Cl^{\ominus}$$

The mechanism is analogous to that in Problem 16–4. At least 2 moles of aniline per mole of acetyl chloride are needed since 1 mole is used to combine with the HCl.

*Problem 16–6.*

$$CH_3CO-O-COCH_3 + CH_3CH_2OH \rightarrow CH_3CO-OCH_2CH_3 + HOCOCH_3$$

Ethyl acetate     Acetic acid

*Problem 16–7.* N-Methylacetamide is hydrolysed by hot alkali.

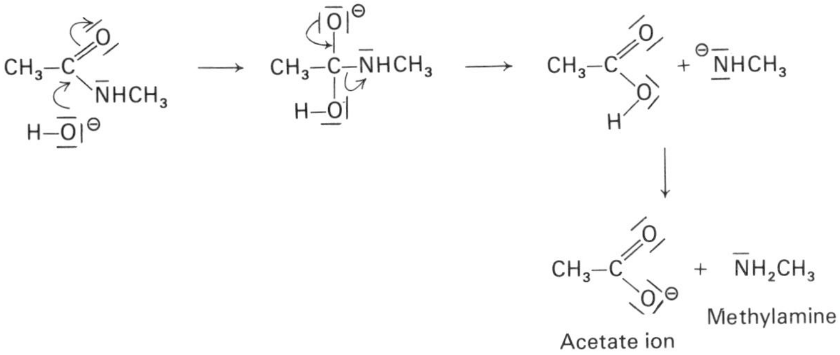

Acetate ion     Methylamine

*Problem 16–8.* If acid catalysis is used, a large excess of acetic acid should be employed so that the concentration of alcohol at equilibrium is small:

$$\text{alcohol} + \text{excess acetic acid} \rightarrow \text{ester} + \text{water}$$

Alternatively, the alcohol could be treated with 1 mole of acetyl chloride or acetic anhydride:

$$ROH + CH_3COCl \rightarrow CH_3COOR + HCl$$

$$ROH + CH_3CO-O-COCH_3 \rightarrow CH_3COOR + HOCOCH_3$$

These reactions do not lead to an equilibrium mixture. Conversion is essentially complete.

*Problem 16–9.* Hydrolysis of the amide with aqueous NaOH would give sodium acetate and cyclohexylamine:

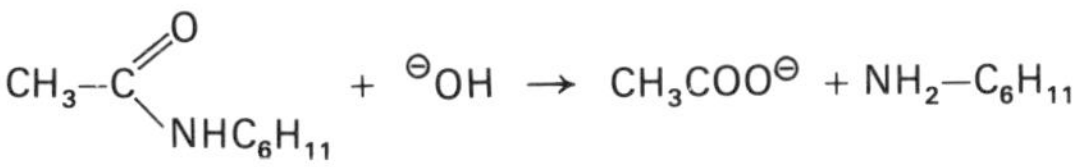

$$CH_3-C\overset{\displaystyle O}{\underset{\displaystyle NHC_6H_{11}}{\big\langle}} + {}^{\ominus}OH \rightarrow CH_3COO^{\ominus} + NH_2-C_6H_{11}$$

The mechanism is analogous to that given in Problem 16–7 for the hydrolysis of N-methylacetamide.

*Problem 16–10.* It is clear that esters could be attacked by many good nucleophiles. Esters react with ammonia to give amides.

$$C_6H_5-\overset{\underset{|}{O}}{\underset{NH_3}{\underset{|}{C}}}-\overline{O}CH_3 \longrightarrow C_6H_5-\overset{\underset{|}{\overline{O}}}{\underset{\oplus NH_3}{\underset{|}{C}}}-O-CH_3 \longrightarrow C_6H_5-\overset{\underset{|}{O}}{\underset{\oplus NH_3}{C}} + ^\ominus |\overline{O}CH_3 \longrightarrow C_6H_5-\overset{O}{\underset{NH_2}{C}} + HOCH_3$$

*Problem 16–11.* In a similar way, ethoxide ion will displace methoxide, forming the ethyl ester. An equilibrium mixture of esters would result:

$$C_6H_5-C\overset{O}{\underset{O-CH_3}{}} + ^\ominus O-C_2H_5 \rightleftharpoons C_6H_5-C\overset{O}{\underset{O-C_2H_5}{}} + ^\ominus OCH_3$$

*Problem 16–12.* The positive charge on the nitrogen will favour proton loss. Nitric acid is a much stronger acid than the hydrogen carbonate ion. The phosphorus atom is larger, so $PH_4^\oplus$ is more acidic than $NH_4^\oplus$. For a similar reason, methanethiol, $CH_3SH$, is more acidic than methanol. Delocalization is possible in the ion $O=N-O^\ominus$, but not in $H_2N-O^\ominus$. Nitrous acid is a stronger acid than hydroxylamine.

*Problem 16–13.* Make use of the greater acidity of the carboxylic acid. Dissolve the mixture in ether and shake with aqueous sodium carbonate. The carboxylic acid will lose a proton to this base and will go into solution in the water as the anion. The phenol will not react and will remain in the ether. Separate the layers. Recover the phenol by evaporating the ether. Recover the benzoic acid by acidifying the alkaline solution with HCl. Benzoic acid will precipitate out.

What would have happened if you had used NaOH in place of $Na_2CO_3$?

*Problem 16–14.*

*Problem 16–15.* The vegetable oil could be hydrolysed with an alkali such as $K_2CO_3$:

$$2\begin{matrix} CH_2-O-CO(CH_2)_{14}CH_3 \\ CH-O-CO(CH_2)_{14}CH_3 \\ CH_2-OCO(CH_2)_{14}CH_3 \end{matrix} + 3K_2CO_3 + 3H_2O \rightarrow 2\begin{matrix} CH_2OH \\ CHOH \\ CH_2OH \end{matrix} + 6KOCO(CH_2)_{14}CH_3 + 3CO_2$$

Ester                  An alcohol     A soap
(glycerol)

The $K_2CO_3$ would be available as a constituent of wood ash.

*Problem 16–16.*

$$R-\overset{|\overset{\ominus}{O}|}{\underset{\underset{H-\overset{\ominus}{A}lH_3}{}}{\overset{||}{C}}}-\overline{O}-CH_3 \longrightarrow R-\overset{\overset{\ominus}{|\overline{O}|}}{\underset{\underset{H}{|}}{\overset{|}{C}}}-\overline{O}-CH_3 \longrightarrow R-C\overset{\nearrow O}{\underset{\searrow H}{}} + {}^{\ominus}|\overline{O}CH_3$$
$$+ AlH_3$$

So far the reaction is analogous to the alkaline hydrolysis of esters. The aluminium hydride will combine with the methoxide ion:

$$AlH_3 + {}^{\ominus}OCH_3 \longrightarrow H-\overset{\overset{H}{|}}{\underset{\underset{H}{|}}{Al}}{}^{\ominus}-O-CH_3$$

All of the hydrogen atoms on the aluminium can be transferred. The aldehyde formed in the first sequence will be rapidly reduced to an alkoxide ion:

$$R-C\overset{\nearrow O}{\underset{\searrow H}{}} + H-\overset{\ominus}{A}lH_2OCH_3 \longrightarrow R-\overset{\overset{O^{\ominus}}{|}}{\underset{\underset{H}{|}}{C}}-H + AlH_2OCH_3$$

which will survive until water is added:

$$RCH_2-O^{\ominus} + H_2O \longrightarrow RCH_2OH + {}^{\ominus}OH$$

The aluminium compounds left will then also decompose, giving $Al(OH)_3$, $H_2$, and methanol. The products of the whole process would be butan-1-ol, methanol, aluminium hydroxide, and perhaps hydrogen if excess of $LiAlH_4$ had been used.

*Problem 16–17.* $CH_3-\overset{\overset{O}{||}}{C}-NH_2$ is not basic because of delocalization.

$NH_2-\overset{\overset{O}{||}}{C}-NH_2$ (urea) is more basic since there are two nitrogen atoms involved in delocalization with the carbonyl group and so the electron density on each nitrogen atom is greater than that on the nitrogen atom of acetamide.

The molecule $CH_3-O-\overset{\overset{O}{||}}{C}-O-CH_3$ is a double ester of carbonic acid. On hydrolysis in acid, methanol and carbon dioxide will be formed.

$Cl-\overset{\overset{O}{||}}{C}-O-C_2H_5$ is a half ester and half acid chloride of carbonic acid. Like other acid chlorides, it will react with amines to give amides:

$$CH_3NH_2 + ClCOOC_2H_5 \longrightarrow CH_3-\overset{\overset{O}{||}}{\underset{\underset{H}{|}}{N}}-C-O-C_2H_5 + HCl$$

$COCl_2$ (phosgene) is the acid chloride of carbonic acid. It reacts with $-OH$ and $-NH_2$ groups in skin or lung tissue, forming esters or amides and liberating corrosive HCl. It is extremely toxic (see p. 82).

# Chapter 17

## A Review

The reactions of organic molecules can be classified in terms of their overall effect as additions, substitutions, eliminations, ionizations, combinations, reductions, oxidations, etc. These are useful descriptions of the net result of the reaction—the difference in structure between reagent and product; but they tell nothing of the mechanism or the conditions by which the conversion can be effected.

For example, the conversion of $R_2C=O$ into $R_2CHOH$ involves the net addition of hydrogen to the ketone. But this could be effected in several ways—by catalytic hydrogenation, by addition of $NaBH_4$ and water, or by treatment with sodium and alcohol. So a net addition could be effected by several different types of reaction mechanism (see Chapter 7).

### CLASSIFICATION OF IONIC REACTIONS

In this book, we have been largely concerned with ionic reactions of organic molecules, that is, reactions in which electrons stay paired and move away from one atom and towards another. If we try to classify ionic reactions we find that, in spite of the variety of molecules and functional groups which can be involved in them, there are relatively few *types* of ionic reaction. The simplest are as follows:

(1) Dissociations such as

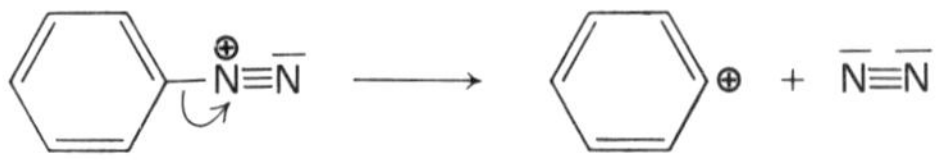

(2) Combinations such as

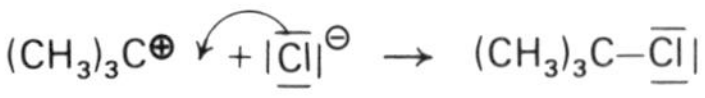

$$(CH_3)_3C^{\oplus} + |\overline{\underline{Cl}}|^{\ominus} \longrightarrow (CH_3)_3C-\overline{\underline{Cl}}|$$

or

$$H^{\ominus} + H_2C\overset{\frown}{=}\underline{O} \longrightarrow CH_3-\overline{\underline{O}}|^{\ominus}$$

180

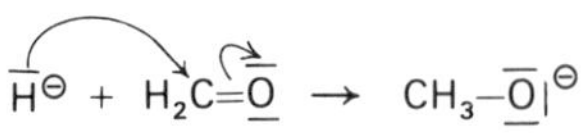

Only slightly more complicated are:
  (3) Substitutions at tetrahedral carbon involving attack by
      (a) nucleophiles or
      (b) electrophiles.
  (4) Substitution at trigonal carbon involving attack by
      (a) nucleophiles or
      (b) electrophiles.
  (5) Addition to unsaturated molecules.
  (6) Elimination, forming unsaturated molecules.
  (7) Substitutions at other atoms, for example hydrogen transfer.
  We shall review each of these reactions in turn, but first we must recognize that all ionic reactions involve nucleophiles and electrophiles.

## NUCLEOPHILES

Nucleophiles are ions or molecules which can donate electrons to form a bond. Make a list of nucleophiles. You will probably start with $HO^{\ominus}$, $NH_3$, $Br^{\ominus}$, and so on, which donate lone-pair electrons to form a bond, but your list could also include benzene, $^{\ominus}BH_4$, $H_2C=O$, $CH_3-Br$ and $CH_3-Li$.

Methyl bromide could be included since it reacts with $FeBr_3$, donating lone-pair electrons to form a bond:

$$CH_3-\overline{Br}| \quad FeBr_3 \longrightarrow CH_3-\overset{\oplus}{Br}-\overset{\ominus}{Fe}Br_3$$

Ethene could be included since it reacts with bromine, donating one pair of bonding electrons from a double bond to form a new bond:

$$CH_2=CH_2 \quad |\overline{Br}-\overline{Br}| \longrightarrow {}^{\oplus}CH_2-CH_2-\overline{Br}| + |\overline{Br}|^{\ominus}$$

Methyllithium could be included since it reacts with formaldehyde (methanal), donating the bonding electrons of a single bond to form a new bond:

$$Li-CH_3 \quad CH_2=O \longrightarrow Li^{\oplus} + CH_3-CH_2-\overline{O}|^{\ominus}$$

Ketones can be included since they react with protons, although it is not possible to tell which electrons are used in forming the bond. Remember that the picture you draw for the product (and therefore the arrows you draw for the mechanism) is not completely adequate.

$$>C=\overline{O} \quad H^{\oplus} \longrightarrow >C=\overset{\oplus}{O}\diagdown_H \left.\right\} \text{ two attempts to represent the same ion}$$

or

$$>C=\overline{O} \quad H^{\oplus} \longrightarrow >\overset{\oplus}{C}-\overline{O}\diagdown_H$$

So we have a varied list of nucleophiles. Some can donate electrons which were originally non-bonding; some can donate electrons which were originally bonding electrons of a double bond; and some can donate electrons which were originally the bonding electrons of a single bond.

## ELECTROPHILES

Electrophiles are ions or molecules which can accept electrons to form a bond. Make a list of electrophiles and classify them according to the manner in which they accept electrons to form a bond. Some, such as metal ions and aluminium chloride, can accept electrons because they have less than eight electrons in the outer shell:

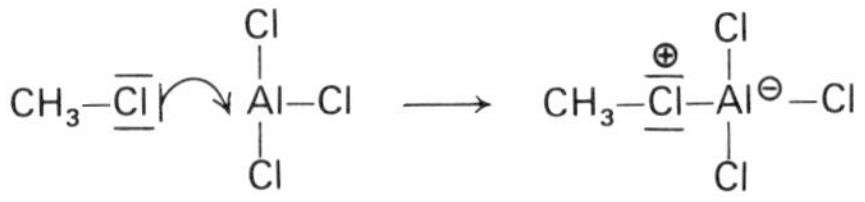

Some, such as methyl iodide, can be attacked by nucleophiles at carbon with breakage of a single bond:

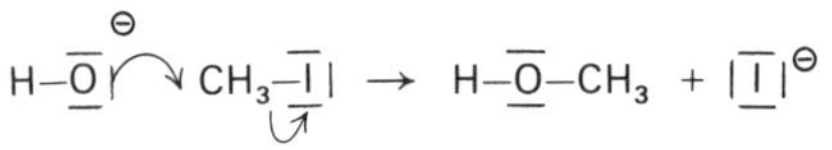

Some, such as the nitronium ion, can be attacked by nucleophiles such as benzene with breakage of one component of a double bond:

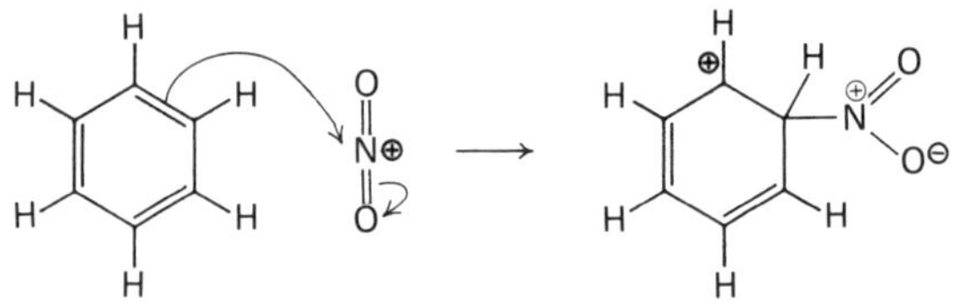

Your list might include $Ag^{\oplus}$, $CH_2=\overset{\oplus}{O}H$, $CO_2$, $CH_3-Br$, $H_2O$, $Br_2$, and $H^{\oplus}$.

Some substances can be replaced in both lists, that is, they can behave either as nucleophiles or electrophiles. Methyl bromide behaves as a nucleophile when treated with $AlCl_3$ but as an electrophile when treated with $HO^{\ominus}$. Formaldehyde behaves as a nucleophile when treated with protons but as an electrophile when treated with methyllithium.

By picking any nucleophile and any electrophile from your lists, you can 'invent' possible reactions. Many of the pairs you pick will react and their reaction may be discussed elsewhere in this book. Other pairs may not react, either because the reaction is reversible and the equilibrium constant is small (i.e. the reactants are more stable than the products), or because the rate of the reaction is very slow.

## IONIC REACTION TYPES

We can now turn to the types of ionic reaction listed above. In the examples given below, and for the first time in this book, lone-pair electrons have been omitted from the mechanistic diagrams. Note that the arrows remain exactly as they would be if the lone pairs of electrons had been drawn in.

$S_N$ *reactions*—substitutions at a tetrahedral carbon atom involving attack by a nucleophile:

$$\overset{\frown}{Y} \quad \underset{\diagup}{\overset{\backslash}{C}}\overset{\frown}{-}X \longrightarrow Y-\overset{\diagup}{\underset{\backslash}{C}} + \overline{X}$$

Examples include the following:

$$H-\overset{\frown}{O}{}^{\ominus} + \overset{\curvearrowleft}{\phantom{.}} CH_3\overset{\frown}{-}I \longrightarrow H-O-CH_3 + I^{\ominus}$$

$$\underset{\underset{H}{\overset{H}{|}}}{H-\overset{H}{\underset{|}{Al}}{}^{\ominus}\overset{\frown}{-}H} + \underset{\overset{\parallel}{O}}{\overset{O}{\overset{\parallel}{CH_3}\overset{\frown}{-}O-\overset{\parallel}{S}-C_6H_5}} \longrightarrow H_3Al + CH_4 + {}^{\ominus}OSO_2C_6H_5$$

$$H\overset{\frown}{O}CH_3 + \underset{\overset{|}{O-CH_3}}{(CH_3)_2C\overset{\oplus}{\underset{\frown}{-}}\overset{\oplus}{O}H_2} \longrightarrow \underset{\overset{|}{O-CH_3}}{(CH_3)_2\overset{H}{\underset{|}{\overset{|\oplus}{C}}}-O-CH_3} + H_2O$$

No attempt is made here to discuss the effect of such reactions on the chirality, if any, of the tetrahedral carbon atom or the timing of the electron pair shifts, except to say that it is possible for the bond formation and bond cleavage to be simultaneous and it is also possible for X to be lost before Y becomes attached to C. You may wish to consider how the structure of the reagents and the reaction conditions might favour one or the other of these two possibilities.

$S_E$ reactions—substitutions at a tetrahedral carbon atom involving attack by an electrophile:

$$Y^{\oplus} \overset{\curvearrowleft}{\phantom{.}} \underset{\diagup}{\overset{\backslash}{C}}{-}X \longrightarrow Y-\overset{\diagup}{\underset{\backslash}{C}} + X^{\oplus}$$

Examples include the following:

$$H^{\oplus} + CH_3-Li \rightarrow CH_4 + Li^{\oplus}$$

$$Br-CH_3 + CR_3-Tl \rightarrow Br^{\ominus} + CH_3-CR_3 + Tl^{\oplus}$$

$$O=CH_2 + CH_3-MgBr \rightarrow {}^{\ominus}O-CH_2-CH_3 + {}^{\oplus}MgBr$$

Again, we shall not discuss the stereochemistry or the timing, except to say that the covalent C–metal bond may break to give a carbanion, which later reacts with the electrophile, or bond breakage and bond formation may be simultaneous.

*Addition–elimination* leading to substitution at a trigonal carbon atom and involving attack by a nucleophile:

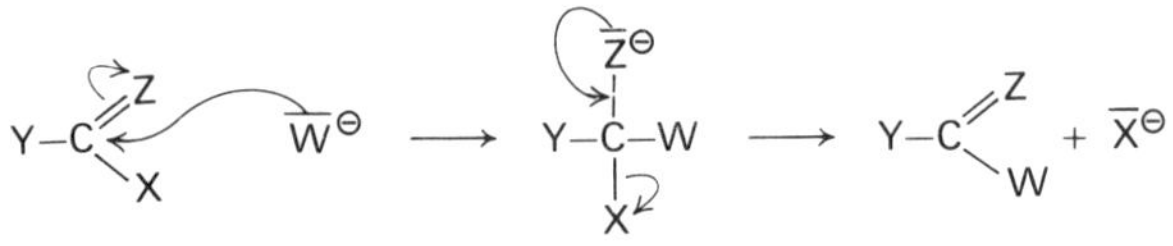

$$Y-C{\overset{Z}{\underset{X}{\Big\langle}}} \quad W^{\ominus} \longrightarrow Y-\overset{Z^{\ominus}}{\underset{X}{\overset{|}{C}}}-W \longrightarrow Y-C{\overset{Z}{\underset{W}{\Big\langle}}} + \overline{X}^{\ominus}$$

Examples include the following:

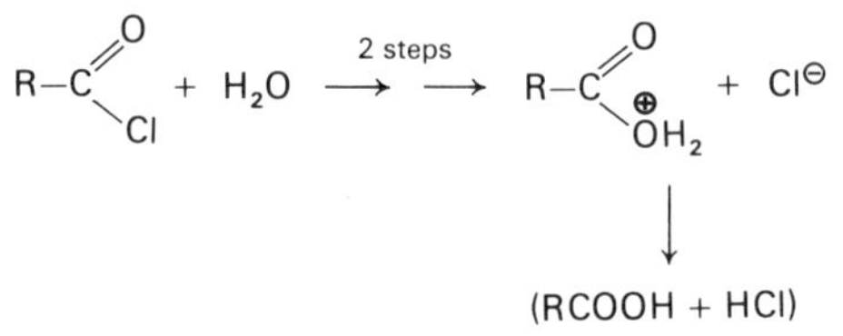

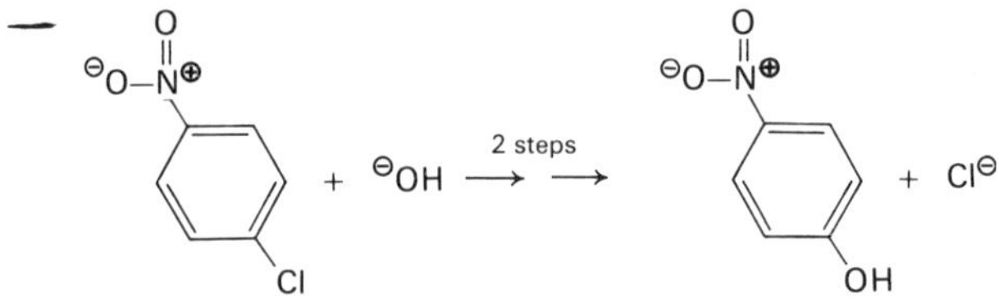

$$R-C{\overset{O}{\underset{Cl}{\Big\langle}}} + H_2O \xrightarrow{\text{2 steps}} R-C{\overset{O}{\underset{\overset{\oplus}{O}H_2}{\Big\langle}}} + Cl^{\ominus}$$

$$\downarrow$$

$$(RCOOH + HCl)$$

$$^{\ominus}O-\overset{\oplus}{N}{\overset{O}{\Big\langle}} \text{—benzene ring—} Cl + {}^{\ominus}OH \xrightarrow{\text{2 steps}} {}^{\ominus}O-\overset{\oplus}{N}{\overset{O}{\Big\langle}} \text{—benzene ring—} OH + Cl^{\ominus}$$

*Addition–elimination* leading to substitution at a trigonal carbon atom and involving attack by an electrophile:

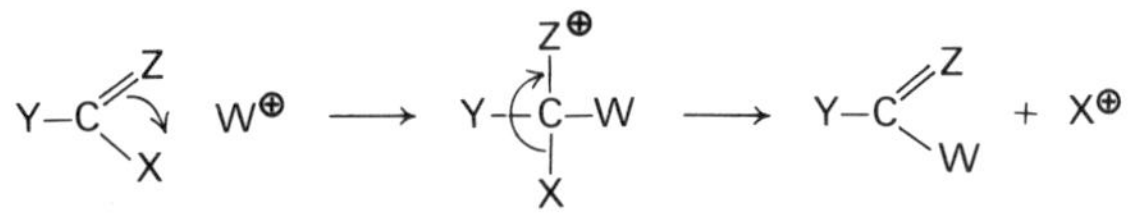

$$Y-C{\overset{Z}{\underset{X}{\Big\langle}}} \quad W^{\oplus} \longrightarrow Y-\overset{Z^{\oplus}}{\underset{X}{\overset{|}{C}}}-W \longrightarrow Y-C{\overset{Z}{\underset{W}{\Big\langle}}} + X^{\oplus}$$

Examples include the nitration of benzene and many other substitutions on benzenoid compounds, including the following case in which $W^{\oplus}$ is a proton and $X^{\oplus}$ a carbonium ion:

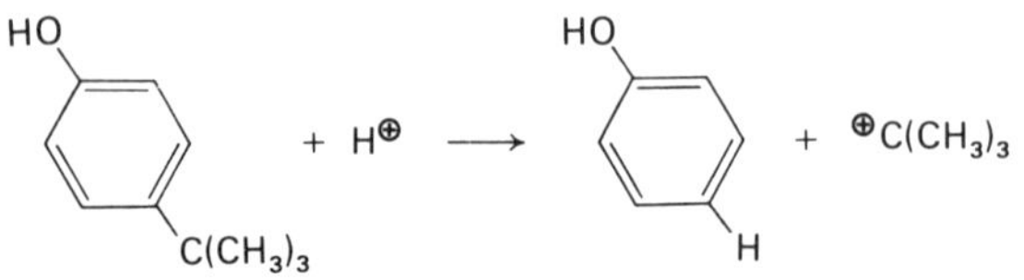

$$HO\text{—benzene ring—}C(CH_3)_3 + H^{\oplus} \longrightarrow HO\text{—benzene ring—}H + {}^{\oplus}C(CH_3)_3$$

In both of these addition–elimination reactions there is again the possibility that the group X comes off before the group W adds. For instance, the acid-catalysed esterification of certain acids proceeds by the following mechanism:

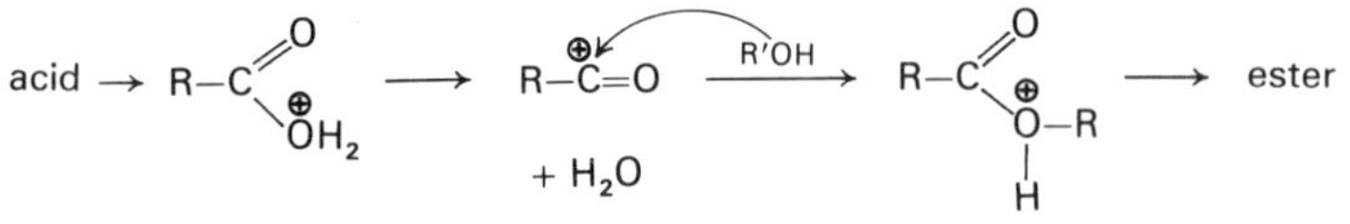

$$\text{acid} \rightarrow R-C{\overset{O}{\underset{\overset{\oplus}{O}H_2}{\Big\langle}}} \longrightarrow R-\overset{\oplus}{C}=O + H_2O \xrightarrow{R'OH} R-C{\overset{O}{\underset{\overset{\oplus}{O}-R}{\Big\langle}}}_{H} \longrightarrow \text{ester}$$

The hydrolysis of diazonium salts also occurs by ionization and then combination:

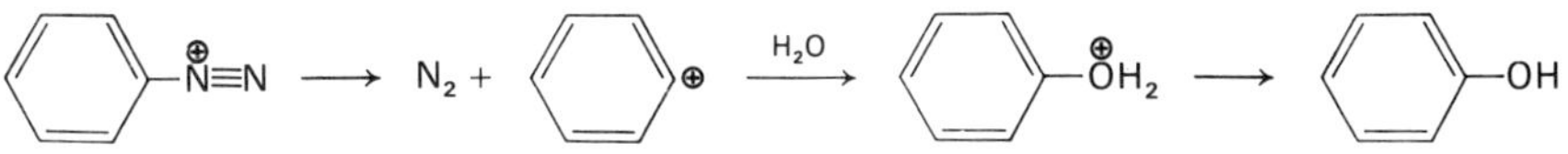

*Additions* to unsaturated molecules such as alkenes and ketones. An electrophilic reagent and a nucleophilic reagent are needed. Depending on the unsaturated substance and the conditions, either may add first. It is very unlikely that they add simultaneously since that would require three independent molecules to collide together. Typical examples include the following:

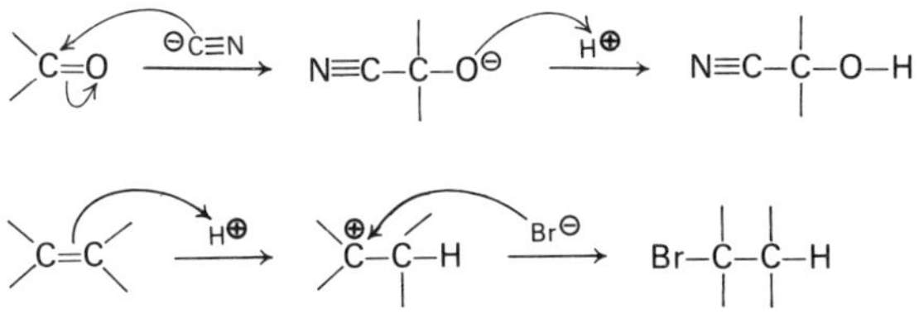

The stereochemistry of the additions has not been discussed.

*Elimination* is the reverse of addition:

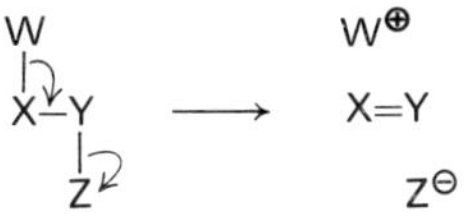

Again, it is possible for $W^{\oplus}$ to be lost first, or for $Z^{\ominus}$ to be lost first, or for the two bond breakages to be simultaneous. These subtleties and the stereochemistry of the reaction cannot be discussed here. Examples include the following:

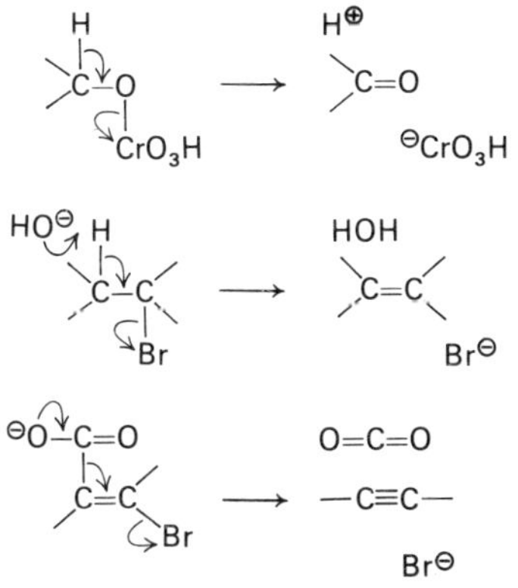

### Substitutions at atoms other than carbon

All of the examples above have been concerned with bonds to carbon, but of course substitutions, additions, and eliminations can occur at other atoms and some cases are discussed elsewhere in this book. For example, the reaction of bromine with ethene involves a substitution at bromine:

$$\text{C=C} \quad Br\text{—}Br \longrightarrow \text{C—C—Br} + Br^{\ominus}$$

and the reaction

$$(CH_3)_3C\text{—}\overset{H}{\underset{}{O}} + H^{\oplus} \longrightarrow (CH_3)_3C^{\oplus} + \overset{H}{\underset{}{O}}\text{—H}$$

is a substitution at oxygen, although we do not usually think of the reaction from that viewpoint.

Substitutions at hydrogen are particularly important. Many of the reactions above include as an almost trivial step the addition of a proton or the loss of a proton from a molecule. The protons have been written as $H^{\oplus}$ but of course free protons do not exist in solutions; they are much too electron-deficient for that. They are solvated and should be written as $H_3O^{\oplus}$ or $H_2\overset{\oplus}{O}R$, etc., depending on the solvent. All of the proton transfer steps therefore really involve transfer of a proton from a reagent or from a protonated solvent and to solvent or to a reagent. Thus, the acidification of sodium benzoate or the second step in the nitration of benzene should more correctly be drawn as follows:

$$\text{(structure)} \quad H\text{—}\overset{H}{\overset{\oplus}{O}}\text{—H} \longrightarrow \text{(structure)} + H_2O$$

$$\text{(structure)} \quad H\text{—}O\text{—}NO_2 \longrightarrow \text{(structure)}\text{—}NO_2 + H_2\overset{\oplus}{O}NO_2$$

These *proton transfers* are thus substitutions at hydrogen involving attack by a nucleophile.

It is also possible to effect substitution at hydrogen involving attack by an electrophile. A familiar example seen in this new light is the reaction of an electrophilic ketone at the hydrogen atom of a tetrahydroaluminate ion displacing aluminium hydride:

$$H\text{—}\overset{H}{\underset{H}{\overset{\ominus}{Al}}}\text{—H} \quad \text{C=O} \longrightarrow H_3Al + H\text{—}C\text{—}O^{\ominus}$$

A new example involves the reaction of carbonium ions with certain alcohols, as occurs in the reaction of diazonium salts with ethanol:

$$\text{(structure)}\text{—}\overset{\oplus}{N}\equiv N \longrightarrow N_2 + \text{(structure)}^{\oplus} \quad H\text{—}\overset{O\text{—}H}{\underset{CH_3}{C}}\text{—H} \longrightarrow \text{(structure)}\text{—H} + \overset{O}{\underset{CH_3}{C}}\text{—H} + H^{\oplus}$$

Benzenediazonium ion

Benzene

Acetaldehyde (ethanal)

Such substitutions can be called *hydride transfers*.

# SUMMARY

The host of ionic reactions of organic molecules can now be correlated. They all belong to one or another of the small number of types listed above. There are no other types of ionic reaction. Within each type there remains the possibility that the electron pair shifts may occur all at once or one after the other. We have not considered these subtleties or the problem of the positions of the atoms as they react (i.e. the geometry of the activated complex); and we have seen very little of the other major categories of reactions—electron transfer, radical and pericyclic reactions.

Nor have we discussed how the mechanisms of reactions are determined, or how knowledge of them allows us to plan controlled syntheses of complex molecules or to understand the reactions of organic molecules that occur in living things rather than in glass flasks.

This has been an introduction to organic chemistry, which is a vast, important, and still developing science. Welcome in!

# Index

Pages in **bold** indicate the first page of a chapter dealing with this topic. Pages followed by the letter P indicate that the topic appears in a problem on that page or in the answer to a problem on that page.

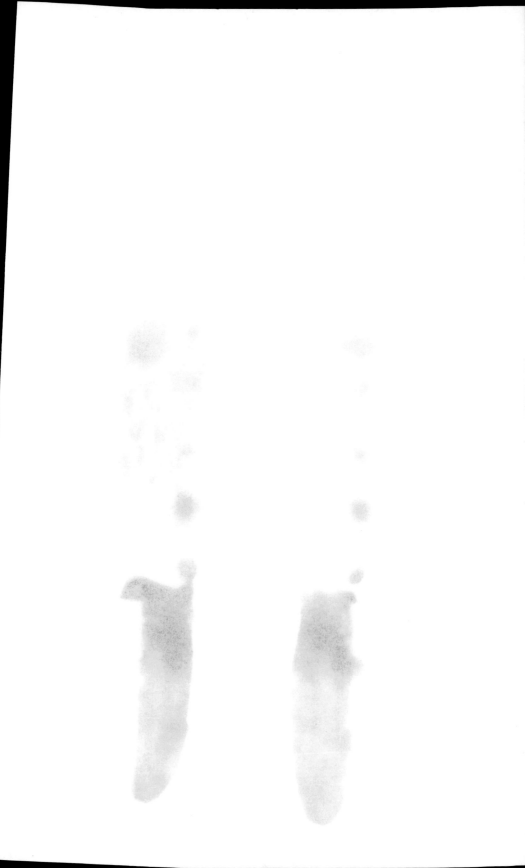

80069

QD
251.2
C37

CARNDUFF, JOHN

AN INTRODUCTION TO

## DATE DUE

| | |
|---|---|
| FEB 08 1995 | |
| APR 2 7 2005 | |
| MAY 0 6 2005 | |
| | |
| | |
| | |
| | |
| | |
| | |
| | |
| | |
| | |
| | |
| | |
| | |
| | |
| | |
| | |

GAYLORD                                    PRINTED IN U.S.A.